I. Minkoff

Materials Processes

A Short Introduction

With 89 Figures

Springer-Verlag
Berlin Heidelberg New York
London Paris Tokyo
Hong Kong Barcelona Budapest

Professor Isaac Minkoff
Department of Materials Engineering
Technion, Israel Institute of Technology,
Haifa 32000, Israel

ISBN 3-540-18895-9 Springer-Verlag Berlin Heidelberg New York
ISBN 0-387-18895-9 Springer-Verlag New York Berlin Heidelberg

Library of Congress Cataloging in Publication Data
Minkoff, I.
Materials processes: a short introduction / Isaac Minkoff.
Includes index.
ISBN 0-387-18895-9
1. Materials, 2. Manufactoring processes. I . Title.

© Springer-Verlag 1992
Printed in United States of America

The use of general descriptive names, registered names, trademarks, etc. in this publication
does not imply, even in the absence of a specific statement, that such names are exempt from
the relevant protective laws and regulations and therefore free for general use.

Typesetting: Laser-Words, Madras, India; Printed in the United States of America
61/3020 - 5 4 3 2 1 0 - Printed on acid-free paper

Foreword

This book is designed to give a short introduction to the field of materials processes for students in the different engineering and physical sciences. It gives an overall treatment of processing and outlines principles and techniques related to the different categories of materials currently employed in technology. It should be used as a first year text and a selection made of the contents to provide a one or two term course. It is not intended to be fully comprehensive but treats major processing topics. In this way, the book has been kept within proportions suitable as an introductory course.

The text has been directed to fundamental aspects of processes applied to metals, ceramics, polymers, glassy materials and composites. An effort has been made to cover as broad a range of processes as possible while keeping the treatment differentiated into clearly defined types. For broader treatments, a comprehensive bibliography directs the student to more specialised texts.

In presenting this overall view of the field of processes, the text has been brought into line with current teaching in the field of materials. The student of engineering, in this way, may see the challenge and the advances made in applying scientific principles to modern processing techniques. This type of presentation may also be the more exciting one.

I. Minkoff

Contents

Chapter 1

Solidification/Liquid State Processes

Solidification is the process of transformation of a liquid to a solid. It is the basis of casting technology, and is also an important feature of a number of other processes including welding, surface alloying, crystal growth, ingot production, materials purification and refining. While it is an important process in metals technology it is also a part of the technologies of ceramics and polymers. In solidification, a solid phase is nucleated and grows with a crystalline structure. For the case where a solid crystalline phase does not nucleate in the cooling process, glassy structures are formed. These represent an important category of material, and processes which involve glasses are included in this chapter.

1.1 Preliminary Concepts

Solidification processes lead to structures which are related to the composition being solidified, to nucleation, to the rate of growth of the solid phase and to temperature gradients in the liquid. The structures can be controlled through these parameters, and variables related to different processes can be changed to effect this control. The structures can be coarse or fine grained. For glasses the structure is amorphous.

A phase diagram for the system being solidified may be used to give a first indication of the structure expected. Figure 1.1, for the Al-Cu system may be taken to demonstrate two possible solidification modes. Between A and B, solidification at ordinary cooling rates would proceed according to a single phase mode. The solid separates from the liquid with composition given by the solidus curve and in equilibrium with liquid compositions given by the liquidus curve. Single phase structures develop and grow, while beyond B a eutectic structure develops at E. These are described in the next section.

Single phase structures may be columnar or equi-axed Figure 1.2 and eutectic structures may be lamellar or rod Figure 1.3. Irregular eutectic structures may also be observed. Finally the structures observed in ternary systems may be somewhat more complex.

A scale may also be defined for the structures which can be defined by a characteristic dimension of the phase, e.g. grain size, or distance between lamellae in a eutectic. Cooling rates, are one of the influencing factors in all solidification

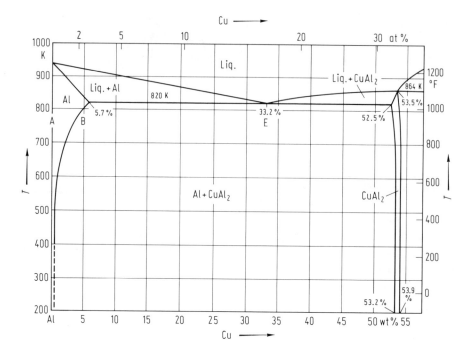

Fig. 1.1 The Al-Cu binary system (from Mondolfo 1976).

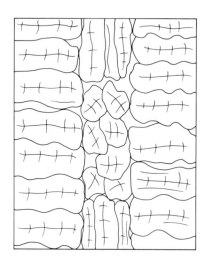

Fig. 1.2 A columnar structure at a casting wall with an equi-axed structure in the interior. These structures are typical of single phase alloys.

processes, and differ according to the thermal conditions imposed. Growth rates of phases are a more important parameter.

A preliminary listing of cooling rates is given in Table 1.1.

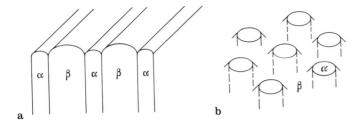

Fig. 1.3 Eutectic structures in most metallic systems are (a) lamellar or (b) rod geometries.

Table 1.1 Cooling rates in solidification processes.

Process	Cooling rate	
Casting	$10–10^2$	Ks^{-1}
Welding	10^2	Ks^{-1}
Gas Laser (Melting)	10^5	Ks^{-1}
Rapid Solidification	$10^5–10^{11}$	Ks^{-1}
Pulsed Laser	$10^{11}–10^{12}$	Ks^{-1}

1.2 Nucleation

The transformation from a liquid to the solid state must be nucleated. At the theoretical solidification temperature T_m, the free energies of liquid and solid are equal, i.e. $F_L = F_S$ and $F_L - F_S = 0$. Below T_m, F_S is smaller than F_L and $(F_S - F_L)$ is a negative quantity. Also the free energy is an extensive quantity and varies with the volume of the phase. Therefore the relationship between $F_S - F_L$ and radius of a growing solid phase looks like the lower curve in Figure 1.4. The quantity $(F_S - F_L)$ has been labelled ΔF_v to relate it to volume. The curve labelled σ is the change of free energy related to the surface energy of the growing solid phase. The middle curve is the total free energy change, ΔF, and is given by Eq. (1.1)

$$\Delta F = -\frac{4}{3}\pi r^3 \Delta F_v + 4\pi r^2 \sigma \tag{1.1}$$

$$\frac{d\Delta F}{dr} = 0 \tag{1.2}$$

$$r^* = -\frac{2\sigma}{\Delta F_v} \tag{1.3}$$

It rises to a maximum at r^* when the maximum free energy change is ΔF^*. The maximum at r^* is given by Eq. (1.2) and r^* is given in Eq. (1.3). This value of r^* is called the critical radius for nucleation, and the curve for the free energy change shows that only when the solid phase radius exceeds this can the solid grow with decreasing free energy. The critical free energy for nucleation, ΔF^* is given by

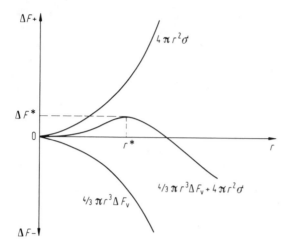

Fig. 1.4 Relationship between the free energy change in a solidification transformation and radius of a growing sphere.

Eq. (1.4). A high value of σ means a high value of r^*, a high ΔF^* and a nucleation problem.

$$\Delta F^* = \frac{16\pi\sigma^3}{3\Delta F_v^2} \tag{1.4}$$

1.3 Heterogeneous Nucleation/Inoculation and other Aspects of Structural Refinement

It is possible to influence the surface energy terms by introducing surfaces into the melt on which nucleation may occur. This may occur naturally at the mould wall. It may also be important that in some processes nucleation at the theoretical temperature is avoided so that the liquid may undercool strongly before solidification, and the growing phase may then develop in a refined manner. One method, which is an important commercial process is by producing small droplets of the liquid, e.g. by atomisation (see Section 1.20).

In normal solidification processes, if a fine structure is required, it is important for nuclei of the new phase to be present at the theoretical temperature of solidification. The nuclei may be added prior to solidification by some process or they may be created in the melt by chemical interaction. Such processes lead to structural refinement. Examples in metallurgical practice are the addition to Aluminium alloys of Ti and B and the addition immediately prior to casting iron of inoculants based on silicon.

There are presently many different methods of structural refinement in casting processes including inoculation, rapid solidification, mechanical or electromagnetic stirring and other techniques.

A short table of advantages of structural refinement is given in Table 1.2. The first two items refer to processes which occur in the liquid state while the third item is of overall importance in achieving superior mechanical properties in the solid state. Hot tearing is a term applied to a separation between metal surfaces in the presence of a liquid phase and is a failure phenomenon accompanying solidification.

1.4 Heterogeneous Nucleation/Theoretical Aspects/Coherent and Incoherent Nucleation

The nucleation of a phase on an existing surface is called heterogeneous nucleation. A new phase (β) appears on a surface in the melt (α) and if the lattice parameters are close enough it adjusts its lattice spacing by an elastic change (strain energy). The difference between the lattice parameters is called the disregistry δ, and δ is given by $\dfrac{\varepsilon_\alpha - \varepsilon_\beta}{\varepsilon_\alpha}$. Figure 1.5 shows a relationship between ΔF^* and δ where ΔF^* is the critical free energy for nucleation, and is related to the disregistry in Eq. (1.6).

$$\delta = \frac{\varepsilon_\alpha - \varepsilon_\beta}{\varepsilon_\alpha} \tag{1.5}$$

$$\Delta F^* = \frac{\mu V^\beta \delta^2}{1 - \nu} \tag{1.6}$$

δ is the disregistry
μ is the shear modulus
ν is Poisson's ratio
V^β is the specific volume of atoms in β phase
ε is the strain

ΔF^* increases as δ increases. A lower value of ΔF^* can be achieved by taking up the mismatch with dislocations, in place of elastic strain; (lower curve).

The interface between nucleated and nucleating phase will now be taken up by dislocations and this is called a semi-coherent interface.

Table 1.2 Advantage of refining cast structure.

---

1. Better mass movement in 2 phase (liquid-solid) state. Better feeding of solidifying metal in conditions of shrinkage.
2. Smaller tendency to hot tearing.
3. Better mechanical properties in as-cast and heat-treated condition.

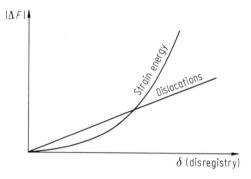

Fig. 1.5 Relationship between the disregistry of a nucleus with the nucleated phase, and the critical free energy for nucleation (from Fine 1964).

1.5 Inoculation in Casting Practice/Aluminium Alloys

The grain refinement of Aluminium Casting alloys may be obtained by the addition of Ti, Zr or V or by the addition of B and Ti together. It is suggested that $TiAl_3$ is the nucleating phase for Ti additions.

1.6 Inoculation of Cast Iron

Inoculation of cast iron is based on ferro-silicon, which may have additions e.g. of strontium. Several models for the inoculation process have been proposed based on the value of lattice spacings in graphite and the spacings for proposed nuclei. These nuclei include SiO_2, spinels and various carbides.

1.7 The Origin of Cast Structure

The discussion of mechanisms leading to the structure observed in cast materials will now be continued. Initially, the subject of nucleation has been presented. In most cases this is heterogeneous, but in some processes of rapid solidification involving atomisation and solidification of small droplets, processes involving homogeneous nucleation may occur.

1.8 Dendrite Growth

Dendritic growth is typical of metallic systems and a representation of a dendrite is illustrated in Figure 1.6. This particular branching geometry has been named after the Greek for a tree and is the result of an instability in growth. A growing solid might be expected to maintain a planar interface and this is true for many cases of growth of crystals but an instability can form under thermal or diffusion influences.

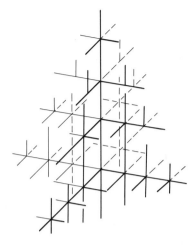

Fig. 1.6 Representation of a dendrite.

Dendritic growth, in the manner of Figure 1.6 dominates the structure of different solidifying systems. It is of course also observed in snow crystal growth from the vapour. In the case of alloy systems, dendritic growth determines the scale of distribution of solute. This is termed chemical inhomogeneity. Castings, weldments, and in general solidified materials are inhomogeneous in that the liquid, and the solid from which it separates, have different compositions and solute inhomogeneity occurs. For the case of a system with a decreasing liquidus temperature solute is enriched in the liquid and the dendritic structure has a varying solute composition across the dendrite arms.

1.9 Transport Processes ahead of a Solid-Liquid Interface

In the solidification process of an alloy, the equilibrium between liquid and solid phases is represented by a common tangent construction to curves of free energy. Figure 1.7 shows the case for a solid solution alloy.

At equilibrium the composition of solid is different from the composition of liquid. On the equilibrium diagram, these compositions at temperature T_1 are shown as C_{S1} and C_{L1}, Figure 1.8a. In this case, the solidifying solid has a smaller solute content than the liquid and the solute accumulates as a boundary layer, Figure 1.8b. The ratio of solute in solid C_s to the solute content in the liquid is termed the distribution coefficient, k.

These solute effects are important in solidification since they lead to segregation, i.e. composition differences in areas which are required to be uniform. While this may be an undesirable effect, segregation can also be used for industrial processes of refining. One process, employed for purifying material, in particular for electronics, is zone refining. Other processes are used in the chemical industry and one important application is in the desalination of water.

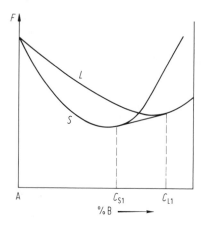

Fig. 1.7 Common tangent construction for equilibrium between liquid and solid phase in solidification of an alloy.

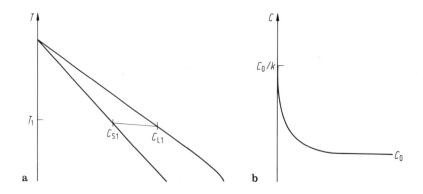

Fig. 1.8 (a) Compositions of phases in equilibrium. (b) Accumulation of solute as boundary layer before a growing solid.

1.10 Microsegregation

Figure 1.9a shows a part of a phase diagram and the variation of solid composition with the fall of temperature, in an alloy of composition C_o as it cools through the solidification range. Figure 1.9b shows this variation through a dendrite arm. From centre to exterior, the composition gradually increases. The spatial composition variation is shown in Figure 1.9c and is roughly sinusoidal. This is a variation in composition on a micro-scale commensurate with the dendrite arm spacing λ. Annealing of castings, e.g. steel, is performed amongst other things to equalise the distribution of solute by diffusion. Since the diffusion time t is a function of the square of distance (in this case λ^2), annealing time is very dependent on cast structure , i.e., to the spacing λ.

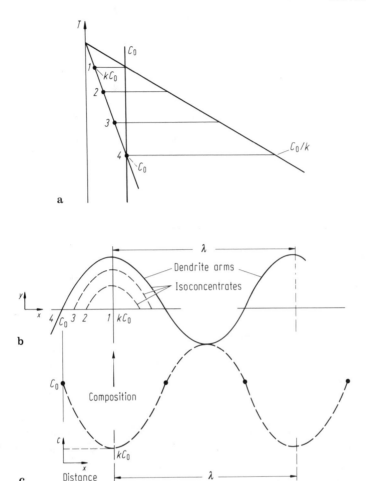

Fig. 1.9 (a) Variation of solid composition in an alloy during the solidification interval. (b) Variation of composition in a dendrite arm after solidification. (After Kattamis, Flemings 1965). (c) Sinusoidal variation in composition of dendrite arm after solidification.

1.11 Zone Refining

The distribution of solute between the solid which separates, and the liquid from which it separated can be used as the basis for a refining operation. The distribution coefficient k, when less than 1, shows that more solute remains in the liquid. In metallurgical processes for producing pure materials e.g. electronics, a molten zone is moved along a bar of the material to be refined, Figure 1.10a. The molten zone can be produced by a High Frequency coil, electron beam, or resistance element.

Figure 1.10b shows schematically the distribution of solute in the bar after a passage of one molten zone. The zone melts off material of composition C_o at

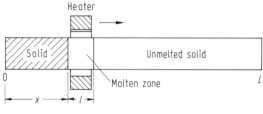

Molten zone of length, l, traversing a cylindrical ingot of length, L

a

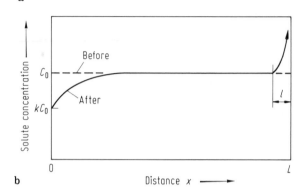

b

Fig. 1.10 (a) Zone refinement process based on movement of a molten zone along a bar (from Pfann 1958). (b) Schematic distribution of solute in bar after passage of one molten zone (from Pfann 1958).

the leading edge and deposits material of composition kC_L at the trailing edge. C_L is the composition of liquid in the zone. Initially this is C_o so that the first solid deposited is kC_o. Gradually the solute content of the liquid zone rises until it reaches C_o/k At this composition the solid deposited is uniformly C_o until the last part to solidify of length l, where the solute content rises.

1.12 Dendrite Growth Theory. Constitutional Supercooling Theory

Figure 1.9a shows part of an equilibrium diagram, and Figure 1.8b shows the distribution of solute in the liquid ahead of an advancing interface. For an initial composition C_o, an advancing solid will be in equilibrium at the interface with a liquid value of C_o/k, where k is the distribution coefficient. Figure 1.11 shows for the advancing interface of Figure 1.8b the conversion of the concentration of solute in the liquid to temperature, T_{CS}. The line G is the actual temperature gradient in the liquid while G_{CS} is the tangent to the temperature T_{CS}. If G_{CS} is greater than G, the shaded area denotes liquid which is below the temperature corresponding to its liquidus. This is termed constitutional supercooling. The concentration of solute in the liquid C_L ahead of an interface moving with velocity R is given by

$$C_L = C_o \left[1 + \frac{(1-k)}{k} e^{-Rx/D} \right] \tag{1.7}$$

D is the diffusion coefficient of solute in liquid.
k is the distribution coefficient

At equilibrium, this is written

$$k_o = \frac{C_s}{C_L} \tag{1.8}$$

The temperature T of the liquidus corresponding to C_L is

$$T = T_o' - mC_o \left[1 + \frac{l-k}{k} e^{-Rx/D} \right] \tag{1.9}$$

The slope of the liquidus curve $m = T/C$.
The temperature gradient in the liquid is

$$\frac{dT}{dx} = T_o - \frac{mC_o}{k} + \mathcal{G}x \tag{1.10}$$

For constitutional supercooling

$$\left| \frac{dT}{dx} \right| \geq \mathcal{G} \tag{1.11}$$

or

$$\frac{\mathcal{G}}{R} \leq \frac{mC_0}{D} \left(\frac{l-k}{k} \right) \tag{1.12}$$

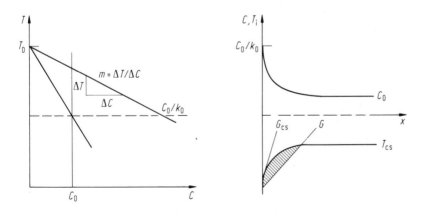

Fig. 1.11 Conversion of the solute concentration ahead of an advancing interface to the equivalent liquidus temperature Tcs. G is the actual temperature gradient ahead of the interface. The shaded area represents undercooled liquid.

Figure 1.12a shows a diagram relating temperature gradient G to R which is termed GR space. This is divided by lines having different G/R values, and which differentiate between planar to cellular instability and cellular to dendritic instability. Cellular instability and dendritic instability are shown in Figure 1.12b.

Branching of the crystal in growth from the primary arm leads to secondary arms and the spacing between these is termed the secondary dendrite arm spacing. This confers a repeat distance on the distribution of solute in cast materials and was described in 1.10. Refinement of the dendrite arm spacing is an important aspect of casting practice. Under ordinary casting conditions this spacing cannot be significantly changed but rapid solidification can lead to a large and important decrease of this value. A relationship between this spacing and cooling is shown in Figure 1.12c. Rapid solidification is discussed in 1.20.

Finally, cast structure in general can be described by the scheme shown in Figure 1.2. This shows a region termed columnar dendritic, growing from the casting wall into a central region termed equi-axed dendritic. In general casting practice with solidification of single phase alloys under conditions of cooling of 10 K per second gives structures which are all columnar or all equiaxed, or they can be mixed, with columnar growth at the walls, and equi-axed growth at the centre.

1.13 Cast Structure

1.13.1 Single Phase Alloys

Single phase alloys are typified by dendritic growth. For the Al-Cu system illustrated in Fig. 1.1, the Cu content of the solid increases towards the boundaries of the growing dendrites. Scheil's equation can be used for predicting composition, for example, to evaluate composition of the solid after a fraction f_s of solidification For the solid

$$C_s = kC_0(1 - f_s)^{k-1} \tag{1.13}$$

For the liquid

$$C_L = C_0 f_L^{k-1} \tag{1.14}$$

C_S is the solid concentration
C_L is the liquid concentration
C_o is the original alloy concentration
k is the equilibrium distribution coefficient
f_L is the fraction of liquid
f_S is the fraction of solid

As an example, of the use of Scheil's Equation, for a 4.5% Cu alloy when 10% of liquid remains, $f_L = 0.1$. For this fraction of liquid and for $C_o = 4.5\%$ Cu,

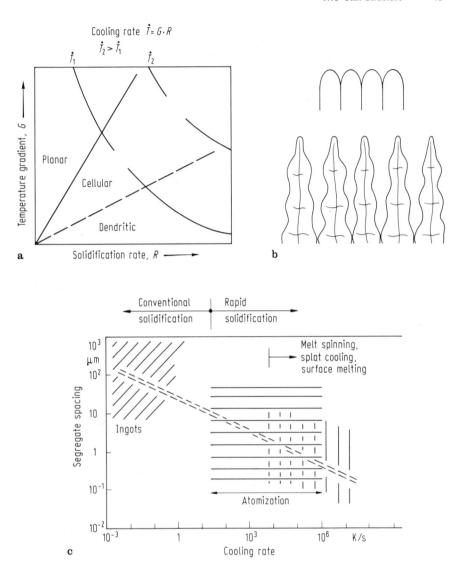

Fig. 1.12 (a) Diagram of GR space relating regions of cellular and dendrite growth to temperature gradient and solidification rate. (b) Cellular and dendrite instability. (c) Relationship between the secondary arm spacing of dendrites (termed segregation spacing) and the cooling rate (from M. Cohen in Mehrabian 1980).

C_L reaches the eutectic composition, 33% Cu. The eutectic should only occur for compositions greater than 5.7% Cu. This demonstrates that castings solidification is non-equilibrium and phases may appear which are not predicted by the equilibrium diagram.

1.13.2 Eutectics

If the liquid composition reaches that of the eutectic a eutectic structure should form by the cooperative growth of two phases e.g. in Al-Cu, by α and $CuAl_2$.

Eutectic structures are common in casting systems, e.g. Al-Si alloys of eutectic composition are in widespread use for light metal pistons. In the Fe-C system, the γ-graphite eutectic constitutes eutectic gray cast iron while the γ-Fe_3C eutectic is the white iron eutectic structure. The eutectic composition has the lowest melting point in an alloy system. Eutectics also have high fluidity. In some systems the two phase structure can have important strength characteristics.

1.14 Eutectic Solidification

Eutectics in metallic systems solidify either as lamellar or rod structures, Figure 1.3. Taking the lamellar structure, the two phases grow cooperatively with a common interface between α and β phases and by diffusion of solute at the eutectic-liquid interface. The solute element B is rejected ahead of the α phase and is incorporated at the β surface.

The solute gradient is the driving force for diffusion and is a function of λ. The velocity of eutectic growth is thus an inverse function of λ. The velocity R and spacing λ are related as follows:

$$\lambda^2 R = \text{constant} \tag{1.15}$$

Eutectics cannot always be strictly regular. The graphite eutectic in cast iron is an example of an irregular eutectic. For irregular eutectics, the exponent can vary, e.g. for gray cast iron

$$\lambda = 0.56 \times 10^{-5} R^{-0.78} \text{ cm} \tag{1.16}$$

The structure of eutectics can be refined by increasing the growth velocity.

There is also a change between the lamellar and rod mode of growth. If S_α is half the width of the α phase and S_β is half the width of the β phase, and $\dfrac{S_\beta}{S_\alpha} = \xi$, then rod growth of α will be favoured for values of ξ given by:

$$\frac{1}{(1+\xi)} = \frac{1}{\pi} \tag{1.18}$$

1.15 Schematic Representation of Casting Processes

A schematic presentation of casting processes is given in Figures 1.13–1.16. These are in effect quite complicated technological procedures requiring skills in control of metal and mould and evaluation of the resulting casting. Mould materials range from sand or plaster to metal. Casting sands require to be refractory to the metal being cast and permeable to gases which originate in filling the mould cavity.

Plaster is specially compounded and used in the lower temperature cast metal applications, particularly Aluminium.

Metal moulds may be of ferrous alloys or copper. Castings may range in weight from grams to 100 tons. Casting may be a mass production process or one of manufacturing individual specialised pieces. Casting alloys must be carefully specified, and in many cases differ from their wrought counterparts, e.g. not all compositions of Al can be successfully cast. There are problems in casting alloys with large solidification ranges or with high shrinkage contraction. Problems encountered in general are shrinkage porosity and cracking.

The processes illustrated are as follows:

Figure 1.13. Sand Casting. The pattern for the casting is moulded in sand and consists of a system for entering the metal into the mould, called "sprue", and

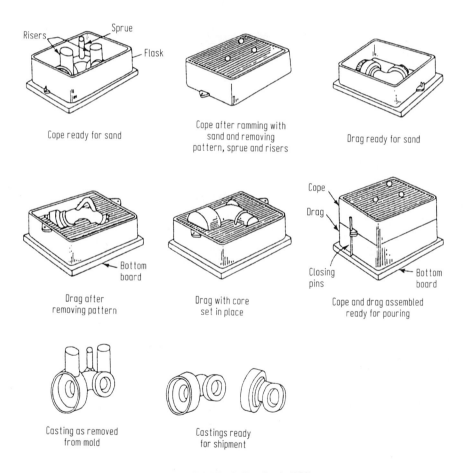

Fig. 1.13 Sand Casting Process (from ASM, Metals Handbook 1988).

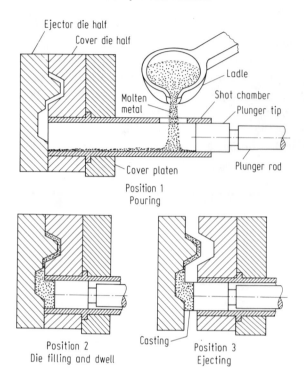

Fig. 1.14 Pressure die casting. Cold die process (from ASM, Metals Handbook 1972).

"risers" which supply liquid metal to compensate for the liquid to solid contraction. The pattern is removed after moulding, the parts of the mould are assembled and the metal is poured into the sprue. The gating system leads the liquid metal into the mould cavity and the risers supply additional metal during solidification. After the casting has been poured and solidified, it is released from the sand, the extra pieces are cut off and after cleaning, additional operations of machining may be performed.

Figure 1.14. Pressure die casting. This process normally used for low melting point metals such as zinc alloys, and Aluminium alloys, uses steel dies. The wall thickness of the castings made is normally in the range up to 4 mm. The illustration shows a cold die process.

Figure 1.15. This process is termed Shell Moulding or sometimes Croning Process. The sand is mixed with a resin binder of the phenolic type and dumped onto the heated pattern. The initial period of polymerisation forms a shell over the pattern. The unaffected sand is returned to the container by inverting the pattern and the shell is released mechanically. It is further heated in a separate cycle to complete the polymerisation process. The different parts of the shell required to complete the mould are assembled and the casting is poured. Steel and cast iron can be cast in these moulds in addition to non-ferrous metals.

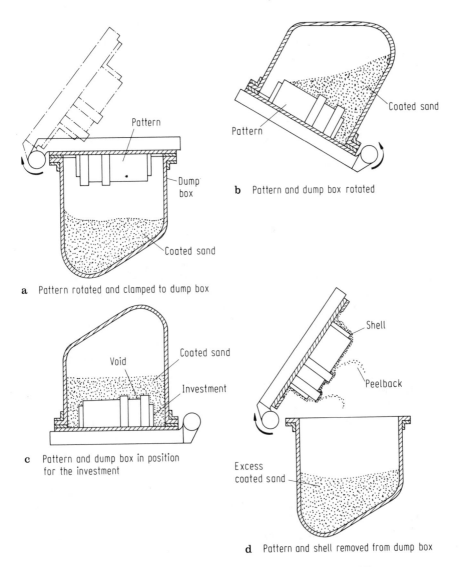

a Pattern rotated and clamped to dump box

b Pattern and dump box rotated

c Pattern and dump box in position for the investment

d Pattern and shell removed from dump box

Fig. 1.15 Shell Moulding and Croning Process (from ASM, Metals Handbook 1972).

Figure 1.16. Lost Wax Process, or Investment Casting. There are a number of variations of this process. The illustration shows the use of wax patterns and a ceramic mould. The wax is injected under pressure into a die to form the pattern for the part to be cast. The patterns are assembled on the sprue into a tree which is then coated by ceramic slurry in several stages. The slurry can be made of SiO_2 particles and the liquid medium is a silicate composition. The initial stage of heating removes the pattern by melting. In the firing stage the remainder of the wax is burnt out of the mould, the ceramic particles are consolidated and the metal is poured while the mould is hot. Processes of this character are also termed precision casting since high accuracy is obtained in dimensions.

1.16 Shrinkage and Risering

In the solidification of castings, the contraction of volume in transformation from the liquid to solid state leads to porosity and loss of mechanical properties, as well as to shape distortion and dimensional change. The problem of porosity will be discussed here. This has two types of distribution determined by the mode of solidification. In materials which solidify over a small temperature interval and in high temperature gradients, the mode of growth is from the mould wall inwards and is termed skin freezing. The shrinkage is located in the central regions of the casting. This is typical of low alloy steels.

For large solidification intervals, and for shallow temperature gradients, the growth occurs over an appreciable area of the casting, where solid occurs together with liquid. This is termed mushy freezing. The shrinkage is then more widely dispersed and is of an interdendritic character.

The latter requires procedures involving care in design, the use of chills, and the correct placement of risers. Steel castings can be approached in riser design by quantitative methods based on riser equations, but preferably by modelling procedures in which the solidification progress is computed and the final areas of freezing located. The location of risers must be determined in relation to the solidification pattern and the risers must be compatible in their geometry with the requirements of supplying liquid metal during the freezing of the casting cross-section. This is described in Chapter 8 on modelling.

1.16.1 Moulds

A mould for making a casting must be designed to accept metal through what is termed a sprue. Figure 1.17. The transfer of metal through runners is termed a "gating" system and risers supply liquid metal to areas where shrinkage is occurring.

1.17 Continuous Casting Processes

It is the practice to cast ingots or long continuous shapes in an uninterrupted (continuous) process. However, for certain requirements and particularly for Aluminium, casting is performed discontinuously (DC process) with lengths cast up to several metres. Figure 1.18a shows a type of continuous casting machine for steel and Figure 1.18b for non-ferrous metals. The growth of the solid wall can be calculated from heat transfer theory and also the shape of the pool between liquid and solid.

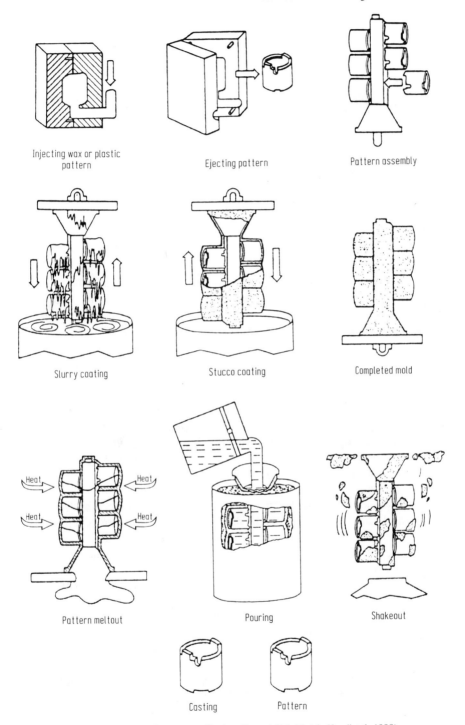

Injecting wax or plastic pattern

Ejecting pattern

Pattern assembly

Slurry coating

Stucco coating

Completed mold

Pattern meltout

Pouring

Shakeout

Casting

Pattern

Fig. 1.16 Lost Wax Process or Investment Casting (from ASM, Metals Handbook 1988).

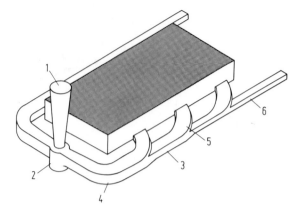

Fig. 1.17 Sprue system for a sand mould.

Continuous casting of steel may be vertical or horizontal. Advantages in continuous casting by a horizontal method lie in equipment construction and in the possibility of one level operation. The development of glass plate production (i.e. Pilkington process) as a horizontal process is discussed in the following.

1.18 Glass Manufacture by the Float Glass (Pilkington) Process

Float glass is a process for making flat glass. In the processes previously employed for making window glass, a sheet was formed by processes which were based on the vertical stretching of a body of molten glass. In drawing from a molten bath, the thickness of the ribbon of glass is controlled by the viscosity. In the vertical process variations in thickness were unavoidable.

The problems were overcome by the float glass process developed by the Pilkington organisation who described the scientific and technical aspects of developing this process. Distortion of glass is problematic, particularly for mirrors or shop windows. The problem could be solved by grinding but this was expensive and wasted 20% of the glass. The Pilkington process was revolutionary in creating glass by a casting process in which glass cools directly to a final distortion free state.

Figure 1.19a shows the float process. A continuous ribbon of glass moves from the melting furnace onto the surface of an enclosed bath of molten tin. A controlled atmosphere shields the tin from chemical reactions, particularly with oxygen, which could result in damage to the ribbon.

The ribbon of glass is held at a high temperature for a sufficiently long time for irregularities to even out and for the surfaces to become flat and parallel. The ribbon is removed by rollers after the cooling process has been sufficient to leave the surfaces sufficiently hard so as not to be marked.

Float glass approaches close to a theoretically perfect process. The influence of surface tension and gravity produce planar surfaces. The width of the strip can

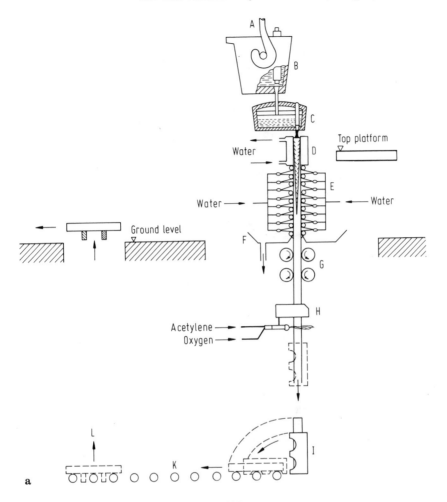

Fig. 18a Continuous casting of steel (from ISI 1977).

be readily varied. The thickness can be varied by controlled stretching in contact with liquid tin. This can reduce the dimension to 3 mm. By arresting the flow it is possible to increase the thickness to 15 mm. An electro-float process was also developed to modify the glass surface by rapid ion replacement between metals and glass.

The thickness of the glass, t, is established by a balance between gravitational and surface tension forces. This is shown in Figure 1.19b. The thickness of the glass is given by the following relationship.

$$t^2 = (\sigma_g + \sigma_{gt} - \sigma_t)\frac{2\rho_t}{g\rho_g(\rho_t - \rho_g)} \qquad (1.19)$$

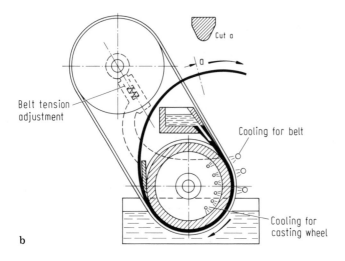

Fig. 18b Continuous casting non-ferrous process (from Emly EF 1976).

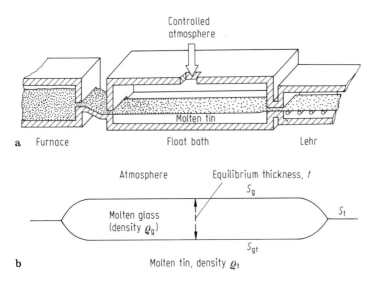

Fig. 1.19 (a) The Pilkington Process for glass (from Pilkington 1969). (b) Thickness of glass established by a balance between gravitational and surface tension forces (from Pilkington 1969).

Here σ is surface tension and ρ is density. The subscripts g, gt and t refer to glass, glass/tin and tin. The g before ρ_g is the gravitational constant.

During the forming process of a normal soda-lime-silica glass, it is necessary to maintain a temperature of 1050 °C, equivalent to a viscosity of $1\ \mathrm{kNsm^{-2}}(10^4 P)$ to produce a surface with a required degree of perfection. A combined gravitational and surface tension force of about 5 Nm^{-2} is available to remove an undulation at the top surface of the glass with an amplitude of 0.1 mm and a wavelength (λ) of 2.5 cm.

At 600 °C, the viscosity is sufficiently high (10 $GNsm^{-2}$ or $10^{11}P$) for the ribbon to be removed mechanically without surface damage.

1.19 Slip Casting of Ceramics

This is a commonly used process for forming ceramics, and makes use of plaster moulds. The ceramic is poured into the mould as a casting "slip", i.e. as a suspension of the ceramic in water. The plaster mould removes water, Figure 1.20, and the resulting body is capable of supporting its own weight.

Deflocculation is practised to reduce the water content of the slip. A normal suspension would settle the particles in "flocks", rather than single particles. In a deflocculated suspension, the particles are individual and give close packing.

1.20 Rapid Solidification

In the previous sections, a description was given of casting processes in which conventionally, cooling rates are 10 °C per sec and the growth rates of solid phases 10^{-2} cm sec^{-1}. These casting processes form the main body of practice for manufacture of cast metal parts in the foundry industry.

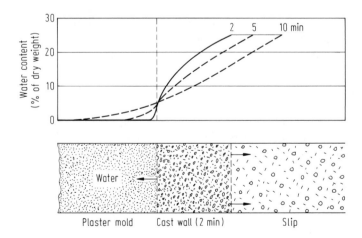

Fig. 1.20 Slip casting of ceramic.

Parts made by conventional castings technology have a cast structure determined by the growth rates of phases and the compositions are given by equilibrium diagrams.

It is possible by suitable methods to solidify metals and materials at very much higher rates, e.g. by forming droplets and thin films. Their cooling rates may approach 10^{11} Ks^{-1}.

This material has some direct application e.g. magnetic materials in strip form may be made this way. It is the practice however in these solidification procedures to further process the rapidly solidified material e.g. by extrusion, as in Aluminium alloy manufacture or by powder metallurgy processes (See Chapter 4).

These methods are now well established e.g. tool manufacture from rapidly solidified tool steel, high strength to density Al alloys e.g. Al-Li, and the manufacture of magnet alloys.

In addition, laser melting and electron beam melting of surfaces can give cooling rates in the metal of 10^5 Ks^{-1} and produce solidified structures of a more refined character than those achieved in ordinary casting.

The overall advantages which can be achieved are related to structure, to extended solid solubility and to the possibilty of obtaining phases not normally present and which are not the equilibrium ones. Amorphous, or glassy structures, can be achieved and these confer both physical and chemical properties which are different from the crystalline phases.

Figure 1.21 shows different methods in rapid solidification processing, which will be explained in the following.

1.20.1 Heat Flow in Rapid Solidification

(a) Heat flow. Figure 1.22 shows the possibilities of heat flow for rapid solidification techniques. In (a) the heat is transferred under Newtonian conditions and can be described by a heat transfer coefficient h and the temperature difference $T - T^A$ between wall and surroundings A_o is the area across which heat is transferred.

In Figure 1.22b, transfer of heat occurs from the growing solid into an undercooled liquid.

The Biot Number $(Bi) = hr/KL$ is a useful parameter for characterising heat transfer. In this, r is a radial coordinate, K_L is the thermal conductivity of the liquid, h is the heat transfer coefficient and L is the latent heat. For high values of (Bi), non-Newtonian heat transfer can occur into the liquid.

For convective heat transfer, h may approach 10^6 Wm^{-2}K^{-1}. For r approximately 500 μm in Fe and Ni, substantial undercooling may occur in the drop and may lead to very high interface velocities in growth.

For the Newtonian case in (a), the average cooling rate is proportional to the film thickness. High cooling rates are obtained with very small film dimensions.

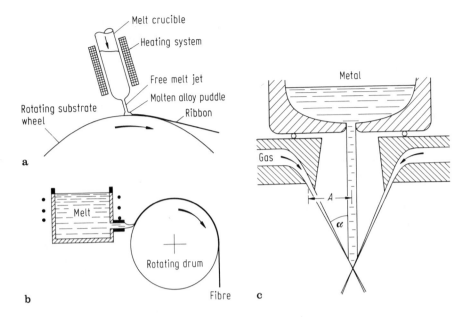

Fig. 1.21 (a) Rapid solidification process. Free jet melt spinning (from Wood 1987). (b) Rapid solidification. Melt drag process (from Wood 1987). (c) Rapid solidification, atomisation (from Wood 1987).

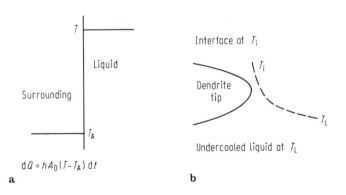

Fig. 1.22 (a) Newtonian heat flow in rapid solidification. (b) Transfer of heat from a growing solid into an undercooled liquid.

1.20.2 Amorphous (Glassy) Structures

Figure 1.23 shows a kinetic model for glassy structures in Salol. The curves are similar to those in solid state transformations occurring in steel and for which TTT diagrams are constructed. These curves are dependent on the nucleation and growth of phases. In Figure 1.23, curve A is for a volume fraction transformed of Salol, $X = 10^{-6}$ and B is for $X = 10^{-8}$. If cooling is carried out to avoid the nose of the curve, it is possible to obtain an amorphous material i.e. no time is allowed for nucleation.

1.20.3 Rapid Solidification and Microstructure

It is convenient to relate secondary arm spacing of dendrites to the different so-lidification processes and note the refinement obtained by rapid solidification. Figure 1.12c shows λ plotted against cooling rate (dT/dt). The structures of in-gots or conventional castings fall in the range of λ approximately equal to 10^3 μm, while melt spinning and atomisation are in the range 10^{-1} μm.

While correlation should be between λ and growth rate of the dendrite, coars-ening processes occur in dendrite growth, and these are related to the time in the solidification interval. As this is reduced in rapid solidification, coarsening becomes limited and the possibility of retaining a refined structure is increased.

1.20.4 Rapid Solidification Processes

1.20.4.1 Atomisation

One type of atomisation processes for producing rapidly solidified droplets was shown in Figure 1.21c. In this Figure the gas stream impinges on the liquid stream in the upper part of the machine. In another type of machine, a gas stream moves round a liquid metal surface from below and takes the liquid stream upwards into an atomisation path.

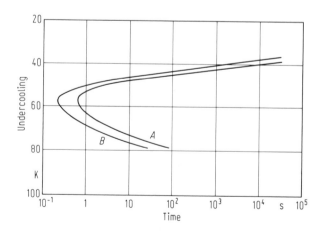

Fig. 1.23 Kinetic model for glassy structure (from Uhlmann 1972).

In chill block melt spinning (CBMS), a stream of molten alloy is brought into contact with a rapidly moving substrate surface. A molten alloy puddle forms on the moving substrate and serves as a local reservoir from which ribbon is continuouly formed and chilled.

Two common CBMS techniques are free jet chill block melt spinning and planar flow casting, Figure 1.21a and b. In the latter the melt is in contact simultaneously with the nozzle and the moving substrate. This delays perturbations of the melt stream and improves strip uniformity.

1.21 Surface Melting

In the first part of this Chapter, the solidification process was described for castings as material in bulk. The conditions of heat transfer through a solidifying wall lead to slow growth kinetics. More rapid rates can be obtained if a film of liquid metal or other material can be created on a solid surface by melting e.g. by a laser. The rate of solidification in the film can then be very much greater than for bulk liquid in a casting and structures determined by conditions of rapid solidification can be achieved. Cooling rates in laser melting are 10^5 Ks^{-1}. Laser processes will be discussed in Chapter 3.

1.22 Osprey Process

The Osprey process uses atomised particles to produce strip or cast shapes, e.g. tubes. Figure 1.24 shows the Osprey process for manufacturing seamless stainless steel tubes. An important technology covered by the Osprey process is the manufacture of billets of composite materials which can be subsequently rolled into sheet or extruded (see next Section).

1.23 Composite Casting

Different methods can be employed for casting metal matrix composites. One of the first approaches to this subject was the directional solidification of eutectic alloys, an important early step in turbine blade manufacture.

Where the composite is based on a dispersed phase, the method of compocasting can be used. This requires the prior mixing of the semi-solid alloy with the phase to be dispersed followed by casting in a mould. A centrifugal casting technique is shown in Figure 1.25. Where the dispersed phase is in the form of fibres, a preform can be employed which is placed in the mould. The metal is introduced between the fibres by pressure, in a technique of the same category as squeeze casting. This is described in Chapter 5.

A further method of casting composites is by the Osprey process in which either ingots, or shapes such as pipe, can be prepared by injecting the solid phase

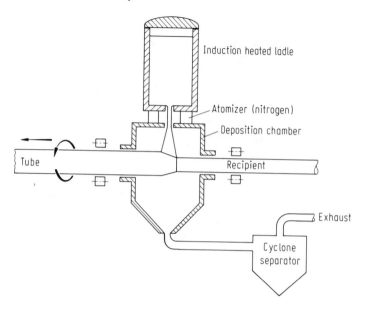

Fig. 1.24 Osprey process.

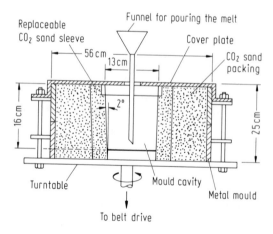

Fig. 1.25 Compocasting process (from Krishnan 1976).

to be dispersed, into the stream of liquid particles projected either into a mould or onto a form. Ingot production of Aluminium alloys having Silicon Carbide as a dispersed phase is now a commercial process, with subsequent shape forming of the ingots by methods such as rolling.

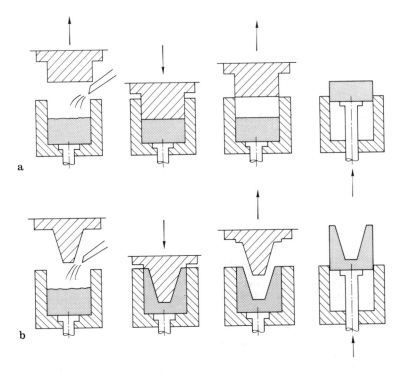

Fig. 1.26 Near net shape process of squeeze casting (from Chadwick GA, Yue TM 1989).

1.24 Near Net Shape Processes

Near net shape processes deal with the problem of producing parts as close as possible to final dimensions. In the casting process, and in particular for sand casting, the dimensions of the part as manufactured are invariably oversize. Thus extensive machining may be required, in addition to the problems of removing risers and the gating system.

Near net shape processes in casting deal with parts conveniently made by a solidification method, using techniques which may be comparable with those e.g. of forging. One process is squeeze casting. This process is appropriate for the manufacture of either ferrous or non-ferrous materials. It employs metal dies and a power press, but starts with a liquid metal which is poured into the lower part of the die, and allowed to partially solidify. The dies are then closed and pressure is applied until solidification is completed. Figure 1.26.

These processes are competitive with forging. They are generally less consuming in energy since little excess metal over the final shape is required. They also have more isotropic mechanical properties.

References

General texts on manufacturing processes

Begeman ML, Amstead BH (1977). Manufacturing Processes, 7th edition, John Wiley, New York.
Bruggeman G, Weiss V (Eds) 1983. Innovations in Materials Processing Plenum Press N.Y.
De Garmo EP, Black JT, Kahser RA (1984). Materials and Processes in Manufacturing, 6th edition, Macmillan.
Erman E, Semiatin SL (Eds) 1987. Physical Modeling of Metalworking Processes, TMS.
Ghosh A, Malik AJ (1986). Manufacturing Science, John Wiley, Chichester
Kalpakjan S (1984). Manufacturing Processes for Engineering Materials. Addison-Wesley.
Lindberg RA (1977). Process and Materials of Manufacturing Allyn and Bacon Inc. Boston, MA.
Murr LE (1984). Industrial Materials Science and Engineering, Marcel Dekker.
Schey JA (1987). Introduction to Manufacturing Processes (2nd edition), McGraw-Hill.
Wang FFY (Series Ed) (1980). Materials Processing, Theory and Practice, Elsevier.
Young JF, Shane RS (eds) 1985. Materials and Processes. Marcel Dekker, New York and Basel.

References on metal casting, solidification and liquid processing

ASM (ed) Metals Handbook, Vo.15 (1988). ASM Metals Park, USA.
Barchfeld FJ, Rostik LF (1987), Near Net Shape Casting ATC, Iron and Steel Society.
Beech J, Jones H (Eds) 1988, Solidification Processing, Inst. of Metals, London.
Biloni H (1983). Solidification. In: Physical Metallurgy, Cahn RW, Haasen P (eds), N. Holland Publ. Co.
Chadwick GA, Yue TM (1989). Principles and Applications of Squeeze Casting, Metals and Materials 5(1):6–12.
Chalmers B (1964), Principles of Solidification, John Wiley, New York.
Davies GJ (1973), Solidification and Casting, Applied Science Publ. London.
Elliott R (1983), Eutectic Solidification Processing of Crystalline and Glassy Alloys, Butterworths, London.
Emly EF (1976). Int Met Rev 21:75.
Evans RW, Leatham AG, Brooks RG (1985), The Osprey Preform Process, Powder Met 28:1 13.
Fine ME (1964). Introduction to Phase Transformations in Condensed Systems, Macmillan, N.Y.
Flemings MC (1974), Solidification Processing, McGraw-Hill, New York.
Geiger GH, Poirier DR (1973), Transport Phenomena in Metallurgy, Addison Wesley.
Jones H (1982). Rapid Solidification of Metals and Alloys. Inst of Metallurgists.
Heine RW, Loper CR, Rosenthal PC (1967), Principles of Metal Casting (2nd edition), McGraw Hill, New York.
ISI (1977). Continuous Casting of Steel. I.S.I., London.
Kattamis TZ, Flemings MC (1965), Trans AIME 1965, 235, 992.
Kondic V (1968), Metallurgical Principles of Founding, Edward Arnold, London.
Krishnan BP, Shetty HR, Rohatgi PK, (1976), Trans AFS 84, 73.
Kurz W, Fisher DJ (1984), Fundamentals of Solidification, Trans. Tech. Publ., Switzerland.
Kurz W, Sahm PR (1975), Gerichte Erstarrte Eutektische Werkstoffe, Springer-Verlag, Berlin.
Mehrabian R, Kear B, Cohen M (Eds) 1980. Rapid Solidification, and Processes, Principles, Technologies. Claitor Pub Div. Baton Rouge LA.
Minkoff I (1986), Solidification and Cast Structure, John Wiley, Chichester.
Minkoff I (1983), The Physical Metallurgy of Cast Iron, John Wiley, Chichester.
Monodolfo LF (1976). Aluminium Alloys: Structure and Properties Butterworths.
Nieswaag H, Schut JW (eds) 1978, Quality Control of Engineering Alloys and the Role of Metal Science, University of Delft.
Pilkington LAB (1969). Proc Roy Soc London, Ser A 314.1.
Pfann WG (1958). Redistribution of Solute During Freezing in Liquid Metals and Solidification ASM.
Ruddle RW (1951), The Solidification of Castings, Inst. of Metals, London.
Uhlmann, DR. J. Non Cryst Solids 1972, 7, 337.
Winegard WC (1964), An Introduction to the Solidification of Metals, Inst. of Metals.
Wood JV (1987). Metals and Materials 3(3):129.
Woodruff DP (1973), The Solid-Liquid Interface, Cambridge University Press.
Zieff M, Wilcox RW (1967), Fractioned Solidification, Edward Arnold, Marcel Dekker.

References on rapid solidification and glassy metals

Ashbrook RL (Ed) (1983), Rapid Solidification Technology, Source Book. ASM.

References on modelling of casting and welding processes

Berry JT, Dantzig J (Eds.) 1984. Modelling of Casting and Welding Processes, Eng. Foundation N.Y.

Brody HD, Apelian D. (Eds) 1981. Modelling of Casting and Welding Processes, Met. Soc AME.

Dantzig JA, Berry JT (1984). Modelling of Casting and Welding Processes.

Fredriksson H (Ed.) 1986. Computer Simulation in Solidification, Proc. E. MRS.

Kou S, Mehrabian R (1986). Modelling and Control of Casting and Welding Processes, Met Soc. Inc. Warrendale, Penn USA.

Markatos NC, Tatchel DG, Cross M, Rhodes N (1986). Numerical Simulation of Fluid Flow and Heat/Mass Transfer Processes, Springer.

Rappaz M (Chairman) 1990. Modelling of Casting, Welding and Advanced Solidification Processes, Enginering Foundation USA.

Sahm PR, Hansen PN (1984). Numerical Solution and Modelling of Casting and Solidification Processes for Foundry and Cast-House CIATF, International Committee of Foundry Technical Associations, 1987.

Smith TJ (Ed) 1987. Modelling the Flow and Solidification of Metals, Martinus Nijhoff.

Chapter 2

Joining Processes

2.1 Scope of Liquid and Solid State Joining Processes

A wide spectrum of techniques embracing liquid and solid state technology is available for the joining of materials. These cover processes for metals, plastics, ceramics, composite materials and packaging in electronics. The welding processes are mainly concerned with the melting of interfaces and employ different sources of energy including arc, plasma, gas, laser and electron beam. In the joining of metals by welding, where additional material is used for forming the molten weld pool, this is obtained from an electrode normally of a composition close to that of the welded metal.

In processes of brazing, a material is melted between the surfaces to be joined and liquid flow by capillarity is necessary to fill the joint. The brazing material is normally different in composition from the brazed material. It is one method for joining ceramics. Bonding by adhesives and solid state processes of joining involving diffusion are important.

2.2 Welding with Gas Sources

The objective in welding processes involving melting is to form a continuous joint across the sections to be joined. Figures 2.1a and b show gas welding. A molten zone is created by melting the edges of both plates and filler metal (electrode). The heat in this process is obtained by the combustion of gas. Typical processes would use oxygen with acetylene, propane, or hydrogen. For acetylene gas, the reaction is:

$$C_2H_2 + O_2 \rightarrow 2CO + H_2 \qquad (2.1)$$

In the envelope, the CO and H_2 undergo further reactions with oxygen:

$$2CO + O_2 = 2CO_2 \qquad (2.2)$$

$$H_2 + \frac{1}{2}O_2 = H_2O \qquad (2.3)$$

The flame provides some protection for metals being joined. A problem in this particular process is the formation of H_2 which is adsorbed by liquid metal. It is for example dissolved readily by liquid Aluminium and hence oxy-acetylene processes would not be recommended for welding Aluminium alloys.

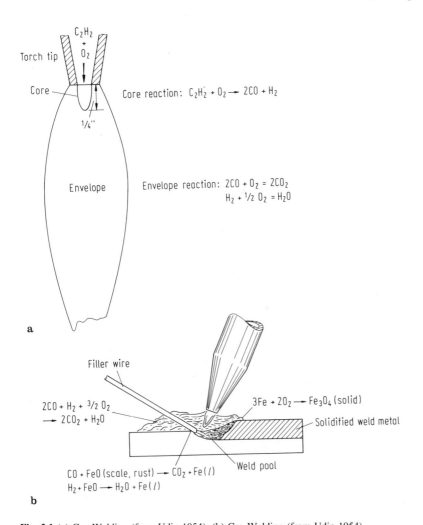

Fig. 2.1 (a) Gas Welding (from Udin 1954). (b) Gas Welding (from Udin 1954).

2.3 Brazing

In the brazing process, materials are joined by melting a braze metal which flows by capillarity into the space between the parts to be joined, and solidifies on cooling. The joint is between solid surfaces via braze material. As an example, in steel brazing, the joint is between the steel surfaces and a copper or copper zinc alloy. Various other compositions may be used and a representative table is shown in Table 2.1. Figure 2.2a shows a schematic arrangement for brazing a cylinder into a hole in a plate. The braze material is in the form of a ring. On melting, it is drawn into the gap.

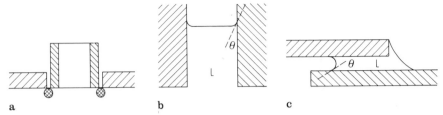

Fig. 2.2 (a) Schematic illustration of brazing process for a cylinder within a plate. A ring of brazing alloy is melted and drawn by capillary forces into the space between the parts. (b) Flow in a vertical joint. (c) Flow in a horizontal joint.

Table 2.1 Brazing materials.

Material composition	M.P.	Application
Cu (pure)	1083 °C	brazing steel in vacuum
45 Cu 35 Zn 20 Ag	790 °C	silver braze (solder)
Ni-Cr-Fe-B	980 °C	high temperature alloys

Brazing is a widespread technique for joining materials, and in particular for difficult cases, e.g. ceramic to ceramic, metal to glass, metal to ceramic.

It may be shown that the strength of a brazed joint in steel is equal or possibly greater than the strength of the steel parts. Its importance as an industrial operation in Metallurgy is related to its employment as a mass production process and to the relatively low joining temperature. Only minimal effects may be noted on the base material. One important application is in joining cutting tool tips to a holder.

The flow of liquid into the space between parts is due to capillarity. The flow may be vertical Figure 2.2b or horizontal Figure 2.2c. L is the liquid phase. For vertical flow, the height reached is $\dfrac{2\gamma}{\rho g D}$, where γ is interfacial energy between liquid and metal, D is the space dimension, ρ is density. For horizontal flow, the distance is $\left[\left(\dfrac{\gamma D}{3\mu}\right)^{t}\right]^{\frac{1}{2}}$ where t is the filling time, μ is the viscosity and γ, D are as before. Therefore, the distance of travel of the filler material may be unlimited. However, vertical flow is preferred because problems are avoided related to the restriction of flow due to air pockets.

2.3.1 Furnace Brazing

In the brazing process, wetting of the surfaces by the liquid braze metal and hence flow by capillary forces is dependent on the cleanliness of the system. A flux is employed to ensure that cleanliness is maintained and wetting occurs.

In furnace brazing, a controlled atmosphere is used with hydrogen or other reducing atmosphere. The brazing materials in electric furnace processes are generally pure copper or copper-silver alloys. Hydrogen does not reduce the oxide of zinc and in the furnace brazing of brass, a flux is required.

Furnace brazing allows a high production rate and the manufacture of assemblies having many joints. These assemblies must be initially located by techniques such as spot welding or pressing. The brazing material is put into the area to be joined in some appropriate form such as wire. For brazing with copper, the furnace temperature is 1140–1150 °C.

2.3.2 Vacuum Furnace Brazing

Brazing operations which require a flux lead to problems in subsequent cleaning since the fluxes are corrosive and must be removed. In these cases, a vacuum can be employed in the brazing process, which allows the surfaces to reach the melting temperature of the braze metal without oxidation occurring.

Vacuum brazing of Aluminium alloys is commonly practised, as is the brazing of stabilised austenitic stainless steels and super alloys. The vacuum employed is about 10^{-5} torr.

2.3.3 Wetting

The surface energies in wetting can be regarded as line tensions and the resolution of forces is shown in Figure 2.3. The angle of contact is θ and the condition for wetting, i.e. for the flow of liquid over the solid is that $\theta = 0$. From Figure 2.3, it is seen that the following condition must obtain

$$\gamma_s > \gamma_{S/L} + \gamma_{L/V} \tag{2.4}$$

i.e. the liquid will wet and run over a surface, if the surface energy of the solid is high, relative to the sum of the solid-liquid and liquid/vapour interfacial energies.

2.3.4 Solders

Solders are low melting point eutectic alloys and are generally used differently from brazing alloys. While the latter flow due to capillarity, solders are applied locally to surfaces. Nevertheless, the wetting condition must still hold. The Pb-

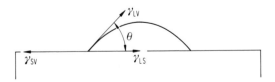

Fig. 2.3 Resolution of forces for a liquid drop on a solid surface (from Udin 1954).

Sn solder will not wet clean solid metals. These solders have a contact angle of 25–70 ° with a steel surface depending on solder composition. However, the tin forms alloys with Fe and so the surfaces are first tinned and the Pb-Sn solder will then wet them.

While Pb-Sn solders have a general application for Fe, the solders employed for Al are Sn-Zn and Zn-Al.

2.4 Adhesive Joints and Wetting

Adhesive joints are made by synthetic resins which solidify by polymerisation. The adhesive bonds operate over small distances only and thus the problem becomes one of wetting of the surfaces by adhesive and filling of asperities. Figure 2.4a shows a substrate which is incompletely wetted and hence incompletely bonded and Figure 2.4b shows one which is completely wetted by adhesive. The problems are in part as discussed for wetting in brazing. In the case of adhesives, the problems are low surface energy of substrate, high polymer viscosity, substrate topography and selective adsorption. A list of adhesives is given in Table 2.2.

Adhesive bonding can be made by a liquid or a solid. If liquid, the adhesive can be either a solvent or non-solvent type. In the former case, the adhesive is dissolved or suspended in a solvent, which must be evaporated before curing is made. Solid adhesives are generally in the form of films or sheets. The adhesive bond is cured at either room temperature or at an elevated temperature by different means including infrared heaters.

2.5 Diffusion Bonding

Diffusion bonding is a solid state process principally employed for joining dissimilar material. Some of the examples given are:- steel bonded to Ti, copper and copper alloys bonded to steel. The process is performed under pressure in a vacuum with heat. A relationship between pressure and bonding temperature can be obtained in which the temperature decreases with increasing pressure. Models for the process are identical with those for the removal of porosity by sintering.

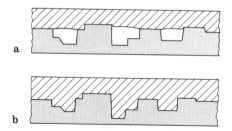

a

b

Fig. 2.4 (a) Incomplete wetting of a substrate in an adhesive joint (from Huntsberger 1965).
(b) Complete wetting in an adhesive joint (from Huntsberger 1965).

(See Chapter 5). The prediction of time to complete bonding is based on porosity removal. This is considered as proceeding in two stages.

1. The flattening of long wave length surface asperities by plastic flow.
2. Reduction of voids by a combination of plastic flow and vacancy diffusion.

Industrial bonding between dissimilar materials in which interlayers are employed, an example being High Speed Steel to 0.45% C steel are basically brazing where a copper layer 0.2 mm thick is used. The temperature of the process is 1073C. The joining of ceramics to metals is also done basically by brazing processes.

2.6 Arc and Resistance Welding Processes

Figure 2.5 shows representative arc and resistance welding processes.

The most common technique is consumable electrode welding using a covered electrode Figure 2.5a. The MIG process Figure 2.5b uses a continuous electrode with a protective gas atmosphere. The gas employed may be Ar, He, or CO_2. Combinations of CO_2 and and O_2 are also used. The MIG process has as a variant, a pulsed arc technique. TIG welding uses a non-consumeable tungsten electrode with an Ar gas protection. (See 2.6.8). Plasma Arc welding is a special type of Arc Welding Process and is shown in (h). Figure 2.5 shows in addition the submerged melt (c), electroslag (d) and electron beam welding processes (g). Figure 2.5(e) shows resistance welding processes.

2.6.1 The Electric Arc

The arc is the most widespread source of heat in welding. It is initiated by the emission of electrons from one surface (cathode) and their acceleration to the anode. The difference in potential between cathode and anode is about 40 V. A plasma of electrons and positive ions is created and heating is effected by the conversion of kinetic energy of the gas.

$1/2mV^2 = 3/2 \text{ kT}$
m is particle mass

Table 2.2 Applications of adhesives.

Adhesive	Applications
Epoxy	Metals
	Ceramics
	Glass
Polyurethane	Coated Abrasives
Polyester	Electronics
Neoprene	Aircraft

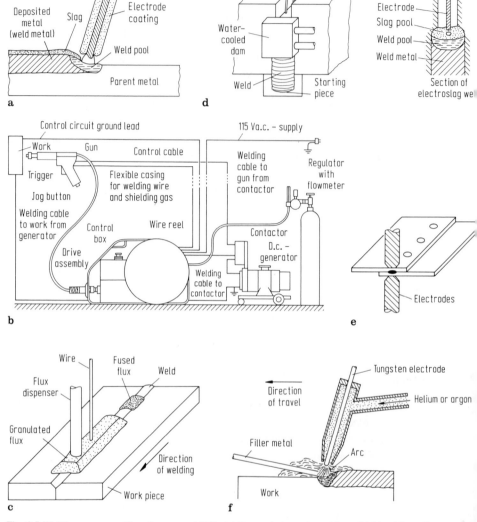

Fig. 2.5 Welding processes (from Lancaster 1965). (a) Covered electrode welding. (b) Metal Inert Gas (MIG) process. (c) Submerged melt welding process. (d) Electroslag process. (e) Resistance welding. (f) Tungsten Inert Gas (TIG) process.

V is particle velocity

k = Boltzmann's Constant

1 eV per particle = 7,500 °C

In arc processes such as covered electrode welding, the arc is started by an initial contact between electrodes which raises the cathode temperature by contact

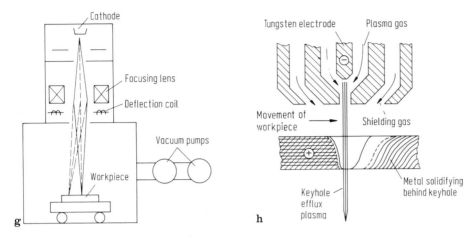

Fig. 2.5 (g) Electron beam welding. (h) Plasma arc welding.

resistance. Electrons are emitted and accelerated in the direction of the positive pole. Gas molecules are converted to ions on impact by electrons having sufficient energy to remove electrons. The positive ions move in the direction of the cathode.

The potential drop across an arc column is divided into three regions, Figure 2.6. About two-thirds of this potential drop is made up of the anode and cathode regions. The anode drop occurs over a length of 10^{-3} cm while the cathode drop occurs over 10^{-3} to 10^{-5} cm.

The ionisation potential of gases and vapours becomes an important quantity in arc welding processes, particularly where protective gases are used.

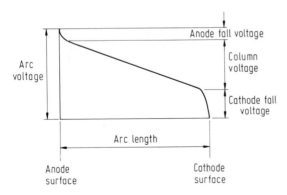

Fig. 2.6 Potential drop across an arc column.

2.6.2 Voltage-Current Characteristics of the Arc

The (VI) voltage-current characteristics of a welding arc are shown in Figure 2.7. The voltage drops initially with increasing current and rises subsequently as for an ohmic conductor. In the figure, the characteristics are shown for an arc in Helium and for Argon at two values of the arc length, (2 mm and 4 mm). The initial fall, and the subsequent rise of the VI curves for the arc show two effects. At low current intensity, the temperature is low and ionisation in the arc is low. With increase of potential, the current intensity increases and the curve is equivalent to that for an ohmic conductor.

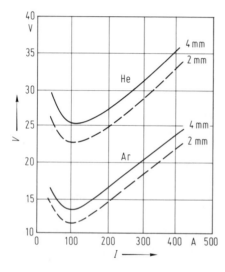

Fig. 2.7 Voltage-current characteristics of a welding arc in He and in Argon for arc lengths of 2 mm and 4 mm.

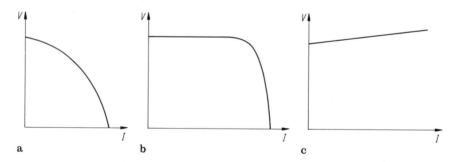

Fig. 2.8 (a) Drooping characteristic. (b) Constant voltage characteristic. (c) Rising characteristic.

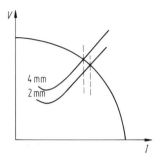

Fig. 2.9 Operating point of an arc for a drooping characteristic.

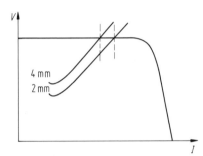

Fig. 2.10 Operating point of an arc for a constant voltage characteristic.

2.6.3 Voltage-Current Characteristics of the Power Supply. Operating Point of the Arc

Different characteristics are imposed by the power supply. Figure 2.8 shows (a) drooping (b) constant voltage (c) rising characteristics. Figure 2.9 shows the arc characteristic related to that of the power source. The operating point of the arc is at the junction of arc and generator characteristics. For a drooping characteristic, which is the type of characteristic generally employed for manual arc welding it is seen that for a change of arc length from 4 mm to 2 mm, the ampere difference is small, so that the heating effects due to control of the arc length manually are not reflected in a large change in diameter of the welding pool.

With a constant voltage arc, Figure 2.10 a small change of arc length is reflected in a larger amperage change than with the drooping characteristic. This type of characteristic is used with automatic welding equipment (e.g. MIG). It controls arc length and stabilises the welding operation.

2.6.4 Covered Electrode Processes

A covered electrode is shown in Figure 2.11.

The coating on the electrode is converted to a slag which protects the liquid particles against oxidation in-flight to the weld. The slag separates, floats to the upper surface of the weld and coats the joint after solidification. It is subsequently removed.

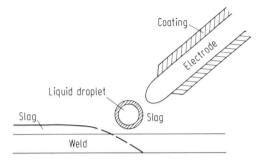

Fig. 2.11 Covered electrode.

Fig. 2.12 Two possible polarity arrangements for welding.

Figure 2.12 shows the two possible polarity arrangements for DC welding. The electrode is negative for straight (or direct) polarity. It is positive for reverse or indirect polarity. The major heating effect is at the anode and direct polarity in DC welding is hence the normal one for covered electrode welding. (Reverse polarity is the common arrangement used for automatic welding). The different polarity arrangemnts in TIG welding using DC will be described.

Table 2.3 gives a short listing of types of coatings for covered electrode welding.

2.6.5 AC Arc

The curves for current and voltage in one cycle of an AC process are shown in Figure 2.13. The current is partially rectified at zero current and positive voltage

Table 2.3 Types of coating for covered electrodes.

ASM number	Coating type	Average composition	Application
E 6010	Cellulose + Na_2SiO_3	50 SiO_2 30 Volatile	Deep penetration
E 6012	Rutile (TiO_2)	30 SiO_2 59 TiO_2	General
E 6015	Basic (Low H)	30 CaO	Low alloy steels. Used when there are problems with hydrogen related cracking

due to the difficulty of the work supplying electrons. At zero current and negative voltage reignition is easy since the electrode is now negative. In some AC welding equipment, a superimposed high frequency voltage reignites the arc and balances the wave.

2.6.6 MIG, Metal Inert Gas Process

MIG process uses a consumeable electrode wire, and a protective gas, i.e. Ar, He, or CO_2. The power supply has either a constant voltage or a rising characteristic. In this process, the desired arc voltage is selected and the feeding rate is adjusted to obtain the proper current. Typical equipment is shown in Figure 2.14.

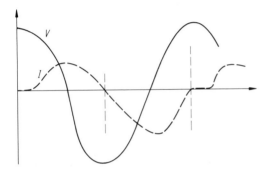

Fig. 2.13 Curves for current and voltage in one cycle of an AC process.

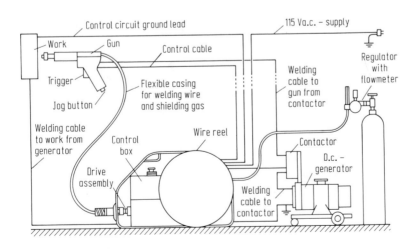

Fig. 2.14 Details of MIG process.

The arc in MIG is self adjusting. the arc length is automatically maintained by the feed rate of the electrode. If the length becomes too small, the amperage rises. This can be seen by studying the curves of Figure 2.7. At any one voltage, the amperage increases as the arc length decreases. This causes the length to increase with the increase in melting rate. If it becomes too big, the amperage decreases, the melting rate decreases and the arc automatically becomes smaller again.

It may be shown that better sensitivity in control of the arc length can be obtained with smaller diameters of feed wire. The variation in current intensity with variation in wire feed rate is dI/dY, and

$$\frac{dI}{dY} = \gamma \frac{F}{K} \tag{2.5}$$

γ = specific gravity of electrode
F = electrode cross section
K = wt. of metal deposited per ampere second
Y = wire feed rate

This shows that refinement in control is dependent on F and hence the use of small electrode diameters in this process.

2.6.7 Metal Transfer in the Arc

The molten tip of the electrode is pinched off and transferred by electro-magnetic (Lorenz) forces to the weld. This is shown in Figure 2.15a, b. Figure 2.15a shows the forces acting on the tip with electrode positive. In the case of electrode negative, cathode spots may be present and these lead to forces acting against droplet transfer, Figure 2.15b. In MIG processes with positive electrode, transfer is not problematic and a spray type of transfer involving droplets of reduced diameter starts at a threshhold current density. This is a desired condition and is shown in Figure 2.16.

The shielding gas has an effect on the burn-off rate. In the MIG process Argon gas can be mixed with O_2 or with CO_2, the former for welding stainless steel and the latter for carbon steel. The burn off rate in this case, at 200 Ampere for carbon steel is twice that for covered electrode welding at the same amperage.

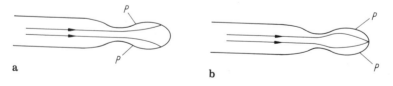

Fig. 2.15 (a) Forces acting on molten tip with electrode positive. (b) Forces acting on molten tip with electrode negative.

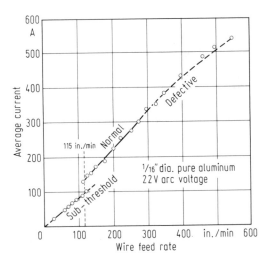

Fig. 2.16 Spray transfer conditions (from Needham, Smith 1958).

2.6.8 Tungsten Inert Gas Welding (TIG)

Replacement of a consumeable electrode by a permanent electrode of Tungsten offers advantages, particularly in employing a protective gas. Alloys of Aluminium are not generally suited to the covered electrode process because of hydrogen adsorption. Stainless steels are problematic because of the high viscosity of Chromium Oxide.

The TIG process is illustrated schematically in Figure 2.17. The arc is struck between a (permanent) W electrode and the work. Argon is used as a protective

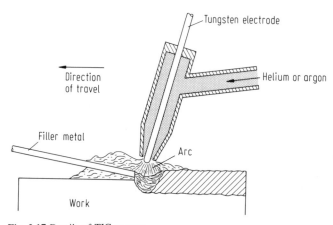

Fig. 2.17 Details of TIG process.

atmosphere. An electrode of the material to be welded may be fed into the weld. In automatic processes, either the work is traversed below the torch or the torch is traversed over the work. DC or AC current may be employed. In the welding of Aluminium and its alloys, it would be useful to use reverse polarity to allow the electron stream to remove oxide from the metal surface. This is not effective thermally because of the high Al conductivity and a compromise solution is the use of AC. The welding of stainless steel is by DC with straight polarity.

2.7 Plasma Welding

A thermal plasma was defined in 2.1 as a gas of electrons and ions in which heat was related to kinetic energy of the particles. In an arc process the electrons are emitted at the cathode and the anode is the work. Alternatively the work may be made the cathode. In plasma welding the process is contained by a chamber. There are two modes, (a) a transferred mode in which the workpiece is the anode as in conventional arc welding, Figure 2.18(a), and (b) a non-transferred mode in which the arc is between an electrode and an annular water cooled copper anode, Figure 2.18(b). A gas stream moves the arc into a constricted region. In both cases, by passing the arc through a nozzle, it is constricted and its velocity and temperature are raised. A power density of about 3×10^{10} Wm^{-2} may be obtained. A property of plasma torches, because of the high gas velocity is deep penetration, e.g. up to 15 mm in stainless steel. The phenomenon of key-holing occurs (also in electron-beam and laser processes) and will be described in 2.9. The power density of 10^7–10^8 Wm^{-2} in a tungsten arc should be compared with the plasma torch power density. For a CO_2 laser, it is 10^{10}–10^{11} Wm^{-2} and for an Electron Beam Process it is 10^{10}–10^{12} Wm^{-2}.

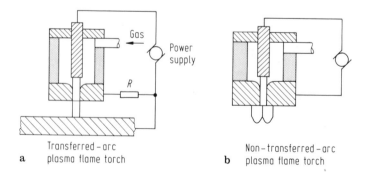

	Transferred – arc plasma flame torch		Non – transferred – arc plasma flame torch
a		b	

Fig. 2.18 (a) Transferred-arc plasma torch. (b) Non-transferred-arc plasma torch.

2.8 Heat Transfer in Welding

Heat transfer theory applied to welding enables the modelling of different parameters related to temperature distribution around a moving heat source. The temperatures achieved are important in assessing the entry of the welded alloy into regions of the phase diagram where transformations occur on heating, and predicting the structures on cooling.

The calculation of cooling rates is used to assess transformations during the cooling part of the cycle. Other important calculations can be made of thermal stresses and the shape and extent of the weld pool and the heat affected zone (HAZ). An area of current interest is to model structure in the solidifying part of the weld and transformations in the HAZ.

The assessment of temperatures in the weld pool is more complex.

The coordinate system used for a moving heat source is shown in Figure 2.19a. The source is taken as origin and ξ is the distance along the x axis from the source which moves with velocity V. Then $\xi = x - Vt$. The strength of the source is q cal s^{-1}. If x, y, z are the coordinates in the stationary system, the equation of heat conduction is

$$\frac{\partial^2 T}{\partial x^2} + \frac{\partial^2 T}{\partial y^2} + \frac{\partial^2 T}{\partial z^2} = \frac{2\lambda\delta T}{\partial t} \tag{2.6}$$

$$\lambda = \frac{\rho c}{2K} \tag{2.7}$$

Substituting for x and differentiating w.r.t. ξ

$$\frac{\partial^2 T}{\partial \xi^2} + \frac{\partial^2 T}{\partial y^2} + \frac{\partial^2 T}{\partial z^2} = -\frac{2\lambda V\delta T}{\partial \xi} + \frac{2\lambda\delta T}{\partial t} \tag{2.8}$$

A steady state distribution of temperature obtains round the source as origin and

$$\frac{\partial T}{\partial t} = 0 \tag{2.9}$$

Then

$$\frac{\partial^2 T}{\partial \xi^2} + \frac{\partial^2 T}{\partial y^2} + \frac{\partial^2 T}{\partial z^2} = -\frac{2\lambda V\partial T}{\partial \xi} \tag{2.10}$$

A solution of this equation can be considered of the following form

$$T = T_0 + e^{-\lambda V}\phi(\xi, y, z) \tag{2.11}$$

T_0 is the initial temperature of the solid and ϕ is a function to be determined. Substituting Eq. (2.11) into Eq. (2.10).

$$\frac{\partial^2 \varphi}{\partial \xi^2} + \frac{\partial^2 \varphi}{\partial y^2} + \frac{\partial^2 \varphi}{\partial z^2} - (\lambda V)^2 \varphi = 0 \tag{2.12}$$

Expressions can be obtained for the distribution of temperature in the welding of thin plates and also thick plates using the geometries shown in Figure 2.19(b) and (c).

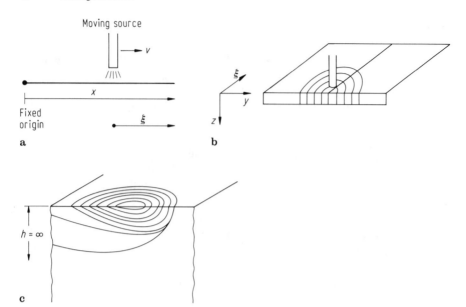

Fig. 2.19 (a) Co-ordinate system for a moving heat source, with the source as origin. (b) Cylindrical distribution of temperature for thin plate. (c) Spherical distribution of temperature for thick plate distorted by movement of source (from Udin 1954).

Figure 2.19(b) for a thin plate, considers a cylindrical distribution of temperature round the source, i.e.

$$\frac{\partial T}{\partial z} = 0 \tag{2.13}$$

Figure 2.19(c) for a thick plate considers a spherical distribution of temperature. For a thin plate, Eq. (2.12) becomes

$$\frac{\partial^2 \varphi}{\partial \xi^2} + \frac{\partial^2 \phi}{\partial y^2} = (\lambda V)^2 \varphi \tag{2.14}$$

The boundary conditions are

$$\frac{\partial T}{\partial \xi} \rightarrow 0 \quad \text{as } \xi \pm \infty \tag{2.15}$$

$$\frac{\partial T}{\partial y} \rightarrow 0 \quad \text{as } y \rightarrow \pm \infty \tag{2.16}$$

If a circle of radius r is drawn round the heat source with

$$r^2 = \xi^2 + y^2 \tag{2.17}$$

then a further condition is

$$-\frac{\partial T}{\partial r} 2\pi r \rightarrow q' \quad \text{as } r \rightarrow 0 \tag{2.18}$$

q' = rate of heat per unit length.

ϕ depends only on r and Eq. (2.12) in cylindrical coordinates becomes

$$\frac{\partial^2 \varphi}{\partial \xi^2} + \frac{1}{r}\frac{\partial \varphi}{\partial r} - (\lambda V)^2 \varphi = 0 \qquad (2.19)$$

The solution of Eq. (2.19) satisfying the boundary condition Eq. (2.18) is a modified Bessel function of the second kind of zero order, $K_0(\lambda v r)$

Hence

$$T - T_0 = \frac{q'}{2\pi K} e^{-\lambda V \xi} K_0(\lambda V r) \qquad (2.20)$$

For three dimensional heat flow in the thick plate, a spherical surface $4\pi R^2$ is taken round the heat source and

$$R^2 = \xi^2 + y^2 + z^2 \qquad (2.21)$$

A solution of the heat transfer equation in 3 dimensions involving ξ and R is

$$T - T_0 = \frac{q'}{4\pi R} e^{-\lambda V \xi} \frac{e^{-\lambda V R}}{R} \qquad (2.22)$$

These equations can be simplified and expressions obtained for the maximum cooling rates at the centre line of the weld.

For a thin plate:

$$\frac{dT}{dt} = \frac{2\pi K \rho c}{\xi^2} \left(\frac{gV}{J}\right)^2 (T - T_0)^3 \qquad (2.23)$$

For a thick plate

$$\frac{dT}{dt} = -\frac{2\pi K V}{\eta J}(T - T_0)^2 \qquad (2.24)$$

g is the plate thickness and only appears in calculations for thin plates.

J is given by the expression $q = \eta J$ where η is the efficiency of heating of an arc and J is the total amount of heat per unit section of plate.

2.9 Electron Beam Processes

Electron beam equipment for welding is shown schematically in Figure 2.20. The electron source is a heated filament or field emitter. Magnetic coils focus the beam and the energy falls within the range 10^{11} W/m^2. The work is held in a vacuum chamber at 10^{-2} Torr. More recent development allows electron beam welding in Argon which considerably extends the scope of the process.

The molten pool in an Electron beam process is shown in Figure 2.21 which illustrates key-holing. The beam travels in a hole surrounded by vapour and this determined the large penetration obtained in the process. Vapour is produced at the initial impact of beam with the surface area and the beam passes through the vapour. Eventually the beam travels in the hole, melting metal ahead and from the sides. A narrow weld zone is characteristic of the process and a narrow HAZ.

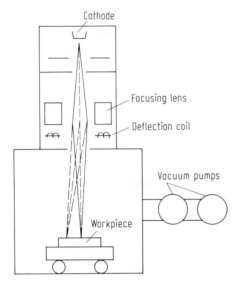

Cathode

Focusing lens

Deflection coil

Vacuum pumps

Workpiece

Fig. 2.20 Electron beam equipment for welding (from Lancaster 1984).

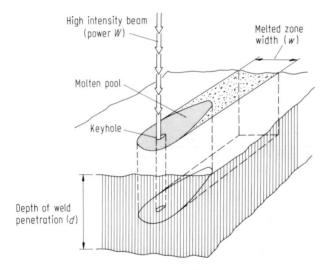

High intensity beam (power W)

Melted zone width (w)

Molten pool

Keyhole

Depth of weld penetration (d)

Fig. 2.21 Molten pool in Electron beam process illustrating key-holing (from Lancaster 1984).

2.10 Weld Structure

The structure of the weld metal is related to solidification. Weld metal structures differ from the structure of ordinary castings because of the following

1. The temperature gradients are higher than in normal casting.

2. The coolng rates and hence growth rates are higher. If dendritic structure alone is considered, the dendrite arm spacing is smaller.
3. Solidification tends to be from the melt/solid interface inwards and into the weld pool. Therefore, the structures are related to initial structures at the interface.

A typical cooling rate calculated for weld metal in arc welding is 10^2 Ks^{-1}. With different types of heat source, e.g. electron beam or laser, and for different conditions of weld pool depth and speed, different cooling rates and different structures result. For the normal welding processes using an arc, the cooling rates lead to only small deviations from the cast structure as described in Chapter 1. Laser heating allows a wider range of structures.

Some types of weld materials offer broad variations of structure related to their composition. Figure 2.22 shows a well-known diagram used in evaluating the structure of stainless steel as a function of Ni and Cr contents. This is called the Schaeffler diagram. At the high Ni side, the structure tends to γ. The ferrite content increases with increasing Cr. The ranges of α, γ and $\gamma + \alpha$ are indicated on the diagram.

2.11 The Heat Affected Zone in Welding

Distribution of temperature round a moving heat source is dependent on the power of the source on the speed of travel, and on the physical properties of the material being welded.

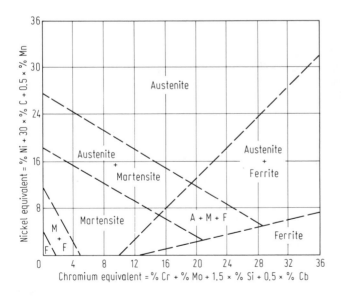

Fig. 2.22 Schaeffler diagram (from Schaeffler 1947).

Figure 2.23 shows a two-dimensional temperature distribution superimposed on a cross-section of the steel weld. The temperature across the weld is correlated with a binary Fe-C phase diagram on which the composition of the weld is taken as 0.15% C.

The structures for steel are related to the heating and cooling cycle. The time at temperature may determine the completion of transformation. The cooling rates determine whether martensite or other structures form. If the cooling rates are sufficient, martensite can form in steel. Both transformation stresses and thermal stresses can lead to cracking.

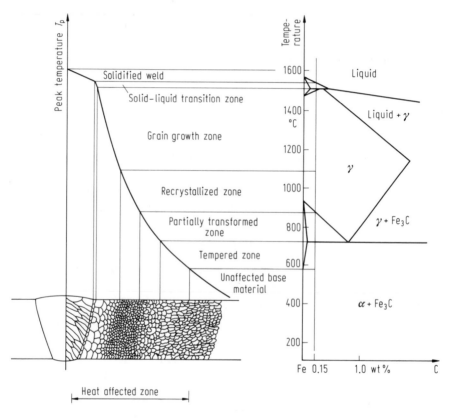

Fig. 2.23 Structure of heat affected zone (HAZ) related to temperature distribution in welding (from Easterling 1983).

References

References on Joining Processes

Adams CM (1958), Cooling Rate and Peak Temperature in Fusion Welding. In: Welding J. Res. Suppl. 37:210–215.

Brown DE, Adams CM (1960), Fusion Zone Structure in Al Alloys. IN: Welding J. Res. Suppl. 39:520–540.

Davies GJ, Garland (1975), Solidification Structures and Properties of Fusion Welds. In: Int. Met. Rev. 20:83.

De Bruyne NA, Houwink R (Eds) (1951). Adhesion and Adhesives Elsevier N.Y.

Easterling K (1983). Introduction to the Physical Metallurgy of Welding. Butterworth.

Garmony C, Paton NE, Argon AS (1975). Diffusion Bonding, Met. Trans. A. 1269–1279.

Houldcroft (977). Welding Process Technology, Cambridge Univ. Press.

Huntsberger JR (1965). The Relationship between Wetting and Adhesion in Advances in Chemistry Series 43.

Kazakov NF (Ed) (1985). Diffusion Bonding of Materials (Transl from Russian) Mir. Publ. Moscow.

Kossovsky R, Glicksman ME (Eds) (1980), Physical Metallurgy of Metal Joining, Met. Soc. AIME.

Lancaster JF (1974). The Metallurgy of Welding, Brazing and Soldering (2nd Ed), Allen and Unwin, London.

Lancaster JF (Ed) (1984). The Physics of Welding, Pergamon, Oxford.

Lane JD (1987). Robotic Welding IFS.

Masubichi K (1980). Analysis of Welded Structures, Pergamon, Oxford.

Rosenthal D (1941). Mathematical Theory of Heat Distribution During Welding and Cutting, Welding J. Res. Suppl. 6:220–234.

Schaeffler AL (1947). Welding Journal 26:601–S

Udin H, Funk ER, Wulff J (1954). Welding for Engineers. John Wiley.

White CW, Peercy PS (Eds) ( ). Laser and Electron Beam Processing of Materials. Academic Press N.Y.

Chapter 3

Surface Processes

3.1 Introduction

The behaviour of manufactured parts is dependent on the structure and properties both of the bulk material and of the surface. Different processes have been developed to treat surfaces and these include diffusion from the gas phase, electrolysis, laser, plasma, ion beam, cladding and thermal treatment. New compositions and new structures can be obtained with important effects on properties such as wear and corrosion. A second category of process dealing with surfaces is related to thin films, particularly in the fields of electronics and optics and these processes involve important applications such as the field of super-conductivity.

3.1.1 Carburising/Carbo-nitriding

One of the first processes used to improve the mechanical properties of steel surfaces was by the diffusion of carbon at the surface into the metal. Originally this was performed by heating with a carbonaceous material. Present methods employ mainly gases, but solid medium (pack-carburising) can be used and liquid carburising has an important place. Plasma nitriding as an important surface treatment process will be discussed in the following section.

These methods are widely used to increase the hardness of the surface of steel parts e.g. gear teeth. For steel parts, surface hardening may be obtained by diffusion of both carbon and nitrogen, into the iron. Hardening by phase transformations is subsequently obtained on cooling.

Reactions in gas carburising mainly involve the gases CO, CO_2, H_2, H_2O, CH_4 and NH_3. A particular reaction might be

$$2CO \rightleftharpoons C + CO_2$$

The reactions lead to carbon dissolved in solid solution in γFe. In carbonitriding, the dissociation of NH_3 provides nitrogen. Carbonitriding is performed at a lower temperature and for a shorter time. The case depth is given as a function of the square root of time $d = \phi\sqrt{t}$. The case and core structures can be varied by heat treatment.

Liquid carburising is performed in cyanide baths. Plasma nitriding, sometimes called ion nitriding, is an important process for building up nitrided surfaces and is discussed in 3.9.1.

3.2 Laser Technology

The treatment of surfaces employs different techniques for heat and mass transfer. The use of lasers will be described in the following.

3.2.1 Types of Laser

Three types of laser used in materials processing will be considered. These are

1. Continuous wave (cw) gas lasers.
2. Solid state lasers.
3. Excimer lasers.

3.2.2 CO_2 Gas (Axial Flow) Laser

Two types of CO_2 gas laser are illustrated in Figures 3.1 and 3.2. Typical power ratings of such lasers for industrial purposes would be in the range of 120 KW. Their engineering applications include marking, cutting, welding, and surface treatment including cladding.

An explanation of the laser can be given as follows: An energy source (which may also be white light shone into the gas), raises atoms through resonance from the ground state to an excited state. In the CO_2 gas laser illustrated, the energy is supplied by a discharge of electricity. Electrons drop back to lower states but many atoms are trapped in a metastable state. A population inversion (i.e. more atoms in a metastable than the ground state) is obtained if the energy intensity is sufficient.

An electron in one of the metastable states can spontaneously jump to the ground state and emit a photon of energy. This can stimulate another atom in the metastable state to radiate a photon of exactly the same frequency and return to its ground state. The stimulated photon has exactly the same frequency, direction and polarisation as the primary photon and exactly the same phase and speed. Both photons can now be considered primary waves and can stimulate other atoms in their metastable states. These emit in the same direction and with the same phase. The primary wave can be absorbed by stimulated transitions from the ground state to the excited states. Hence an excess of stimulated emission requires more atoms in metastable than in ground states. These processes lead to high intensity coherent radiation.

The stimulated emission must be collimated by designing a proper resonant cavity in which the waves are re-used. If the gas is contained in a cylindrical vessel, two highly reflecting mirrors are placed at the cylinder ends as seen in the illustrations and photons emitted parallel to the axis are reflected between the mirrors. The chance of these photons stimulating emission depends on high reflectance of the end mirrors and high population density of metastable atoms within the vessel. Lasing occurs when the build up of photons moving back and forwards becomes self-sustaining. The system then oscillates spontaneously. The laser system therefore consists of

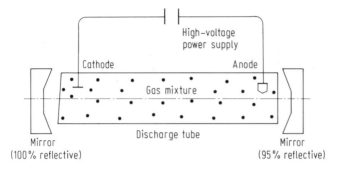

Fig. 3.1 Axial flow type gas laser showing tube.

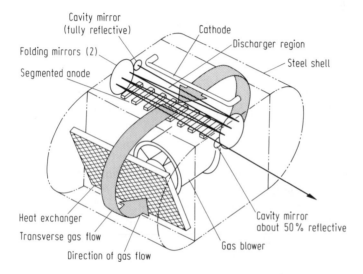

Fig. 3.2 Transverse gas flow laser.

1. A material or medium that is capable of maintining a population inversion of metastable atoms.
2. A method of producing and maintaining the population inversion.
3. A collimation system that reflects the waves back and forth in the cavity.

   In the axial flow type gas laser which is partly illustrated in Figure 3.1 The gas flow is in the same direction as the laser beam. The axial flow of gas is maintained through the tube to replenish molecules depleted by the multikilowatt discharge of electricity used for excitation. Figure 3.2 illustrates a transverse gas flow laser.

   This operates by continuous circulation of the gas across the resonator cavity by a high speed blower. The electric field is maintained perpendicular to both the gas and the laser beam. The volume of the resonator is large relative to its length, so that large mirrors can be placed at each end to reflect the beam through the discharge region several times before it emerges through the output coupler.

3.2.4 Solid State Laser

Figure 3.3 shows a ruby laser. The ruby is in the form of a cylindrical rod with the ends polished flat and parallel. Excitation is by means of intense Krypton or Xenon optical lamps. The laser can be continuous or pulsed. The continuous laser uses Xenon lamps (up to 10 W) and Krypton (up to 100 W).

3.2.5 Excimer Laser

Excimer lasers use rare gas halides e.g. Krypton Fluoride. These are pulsed lasers in which stimulated emission in rare gas halides is obtained by methods such as an electron beam, or by an ultra-violet pre-ionised discharge. The ArF excimer has an operating wavelength of 193 nm.

A typical system with an electron beam source consists of a high voltage generator, a pulse forming line to produce ideally square pulses and a vacuum diode. In the pre-ionised discharge apparatus, an initial electron density of 10 electron/cm is obtained by corona discharge or by an X-ray source.

Excimer lasers find interesting applications in electronics and also in medicine. The interesting property of the ultra-violet light from these sources is in the number

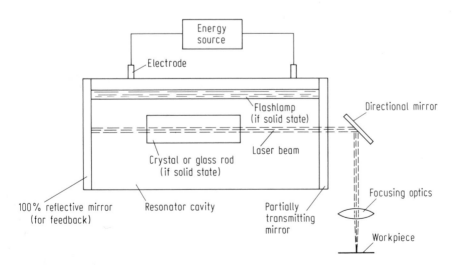

Fig. 3.3 Solid state laser.

of photons per cm^2 and second. For an organic material, e.g. a polymer, or for a biological material, the laser light is absorbed and then the bonds in large molecules are broken. These produce a high concentration of volatile molecular fragments. The pieces ablate and carry most of the excess energy. Hence little heating effects are noted in the unirradiated parts of the material.

3.3 Transformation Hardening of Surfaces by Laser

Solid state hardening (Austenite to Martensite) of steel, or cast iron by a laser is a growing practice. Induction hardening and flame hardening are employed in a wide variety of applications but the use of laser hardening is a more recent innovation. It has advantages over previous practice in being capable of control over very small dimensions in area and in depth. A typical application would be hardening of valve seatings in engines. A growing application is in the hardening of cylinder barrel inserts. The action of the laser is to transform the material by heating into the γ region, followed by cooling at rates which lead to different transformation products.

Figure 3.4 shows the geometry for a laser arrangement which heat treats the surface of steel. In the process illustrated a laser of up to 2.5 kW power was used with beam diameter obtained by focusing between 1–10 mm, and beam velocities

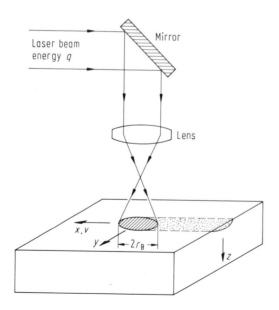

Fig. 3.4 Geometry of a laser arrangement for heat treating surface of a steel (from Ashby, Easterling 1984).

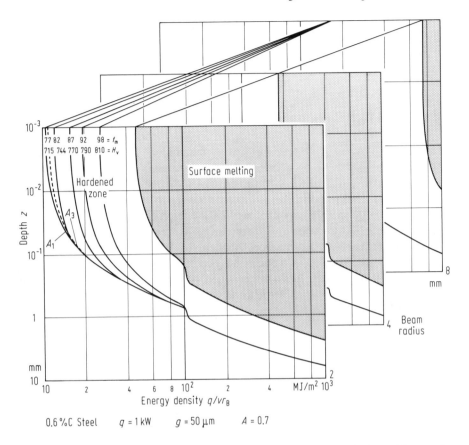

Fig. 3.5 Hardness values in laser surface process on steel related to energy density and beam radius (from Ashby, Easterling 1984).

in the range 2–30 mms^{-1}. The energy density was 10–100 MJm^{-3}. Figure 3.5 shows a diagram, which relates hardness values at depths below the surface (z), energy density, and beam radius.

In the laser hardening of steel, transformation products vary from the surface inwards both in structure e.g. ferrite and martensite, and in hardness. The variation of martensite hardness from the surface inwards is due to the increase of C content with depth.

3.4 Laser Annealing/Laser Annealing of Silicon

Lasers have been effectively used as a heating source for annealing of surfaces in addition to the other functions such as hardening, melting, alloying.

One such annealing application is in the field of micro-electronics, after impurity doping of Si by ion implantation. The dopant impurity varies in its distribution and damage occurs in the Si which may overwhelm the doping characteristics of the impurity. Both continuous wave and pulsed lasers may be used for annealing.

3.5 Laser Alloying

Lasers are employed for alloying surfaces by melting and adding alloy elements. Different methods may be used, including predeposition of the elements by painting, followed by drying of the surface and then diffusion. Electroplating may be employed. Powder injection into the beam as well as powder dispensation ahead of the traversing beam, are important surface alloying processes.

Important applications are to be found in tele-communications. An example is the alloying of Al to Ni at a Ni surface. Other applications e.g., corrosion protection and heat resistance coatings are obtained by surface alloying of materials which include Cr and Mn. There is a growing field of application in applying boride and carbide surfaces on metals. Particles may be injected into the laser beam, e.g. Boron Carbide, to give hard surfaces on steel.

3.6 Cladding

3.6.1 Explosive Cladding

The cladding process on metals produces a layer which may offer corrosive protection in harmful (chemical) environments, a hard metal layer for wear protection, and other applications. An explosive is used and the cladding material is impelled against the parent plate. The bonding involves different physical mechanisms which will be described.

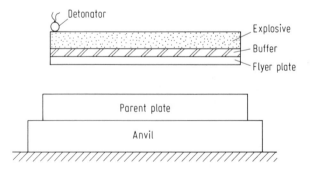

Fig. 3.6 Explosive cladding using parallel technique (from Ganin 1984).

Two techniques are illustrated in Figure 3.6 and Figure 3.7. These are (a) a parallel technique and (b) a technique in which the plates commence at an angle. There is no basic difference but some advantage is obtained in quality of the product.

In Figures 3.6 and 3.7a the layer of explosive is shown covering the upper plate. The explosion starts at one end and travels in the direction AC. The geometry of the plates during cladding in both cases is shown in Figure 3.7b. The upper plate is called the "flyer". Figure 3.7b shows the geometry during cladding for plates originally at an angle α. The flyer makes an angle β with the parent plate. A schematic illustration of the flow of material in cladding at impact is shown in Figure 3.7b. A jet of material is released in the impact area and the interface adopts a sinusoidal wave geometry. The melting of metal at the interface is a factor in the joining mechanism.

The mechanism of explosive welding has been considered to relate to this jetting process. The process is likened to the collapse of a conical or wedge-shaped liner in a hollow charge. A high velocity jet is formed which has high powers of penetration. The kinetic energy of the jet is converted to thermal energy and some melting will occur.

The important parameters in the process are the detonation velocity of the explosive (V_D) and the velocity imparted to the flyer plate (V_p). The velocity of the flyer plate relative to the point of contact at the weld(s) must be less than V_D i.e. $V_p \tan \beta < V_D$ for jetting to occur.

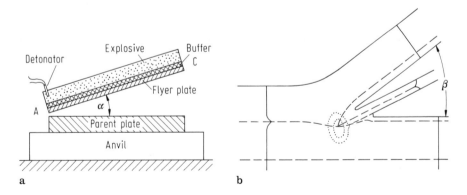

a   b

Fig. 3.7 (a) Explosive cladding with plates at an angle (from Ganin 1984). (b) Geometry of plates during cladding. Schematic illustration of flow of material showing formation of a metal jet (from Birkhoff 1948).

3.7 Coating

3.7.1 Evaporative Coating

Different energy sources may be used for achieving evaporation of material in order to coat a surface. These may include the following:

a. Resistance Heating
b. Electron Beam Heating
c. Hollow Cathode Evaporation
d. Radio frequency induction heating
e. Sputtering
f. Laser heating
g. Plasma

Techniques involving sputtering are described in the following section.

3.7.2 Sputtering

This is a term used for evaporating a source of material to be used as a coating employing an electric discharge. The process is performed in vacuum. An advantage of the sputtering process is considered to be its generality since any material which can be treated in a vacuum can be coated.

The process operates by an electric discharge in an inert gas with the target and the substrate acting as electrodes. Positive ions created by the discharge are accelerated towards the cathode. Material sputtered from the cathode is mostly in the form of neutral atoms together with particles in the form of ions. The particles condense on the substrate. The discharge is both dc and rf. For insulating target materials, a surface charge is built up and must be neutralised. The substrate is bombarded by inert gas molecules, by electrons, photons and negative ions from the plasma, in addition to the desired atoms from the target.

Figure 3.8a shows a simple cathode sputtering apparatus. The electrical discharge is initiated between two parallel plate electrodes, one of which is the target and the other the substrate.

The number of atoms per unit area leaving the target is given by

$$N = \frac{J_+}{g} S(V, A, B) \tag{3.1}$$

J_+ is the current density of the bombarding ions
g is the number of electron charges per ion
S is the sputter yield of atoms per incident ion
    (S is a function of V, the ion energy, A, the ion species, and
    B, the target material.)

The material to be deposited is made the cathode and the substrate is made the anode. The voltage applied is 10–15 kV. The apparatus contains inert gas, either Ar or Xe at 10^{-1} to 10^{-2} Torr.

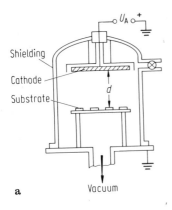

Shielding

Cathode

Substrate

U_A +

d

a Vacuum

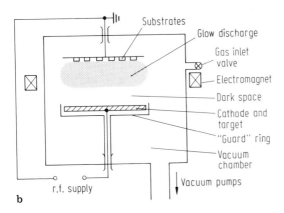

Substrates

Glow discharge

Gas inlet valve

Electromagnet

Dark space

Cathode and target

"Guard" ring

Vacuum chamber

Vacuum pumps

r.f. supply

b

Fig. 3.8 (a) Cathode Sputtering Apparatus. (b) HF process. One electrode is coupled to an rf generator.

A glowing plasma is formed, separated from each electrode by a dark space, Figure 3.9. Electrons are drawn out of the plasma boundary close to the anode and accelerated across the anode dark space. Ions are drawn out of the other boundary and accelerated towards the cathode.

Electrons generated at the cathode surface collide with neutral molecules. The electron-ion pairs replenish the plasma.

3.7.3 High Frequency AC Discharge

HF, AC discharges of 5 to 30 MH_z are employed. In the case of the DC process described in the previous section, problems are encountered in initiating the process, and with sputtering insulating materials. In the HF process, one electrode is coupled capacitatively to an rf generator, Figure 3.8b. It then develops a negative DC bias with respect to the other electrode. This is related to the easier response of electrons than ions, to an applied rf field.

3.7.4 Reactive Sputtering

One form of deposition by sputtering is called reactive sputtering. The target can be a pure metal, an alloy or a combination of materials. The sputtering is performed in a pure reactive gas, which is eventually to be a component of the film, or a mixture of inert and reactive gases.

Figure 3.9 shows DC planar magnetron sputtering. Magnetrons are further discussed in Section 3.10. This type of equipment can be used for sputtering Titanium Nitride on a surface at 5×10^{-7} Torr.

As an example Ti is sputtered from a 6'' target in a mixture of Ar and N_2. The flow rate of N_2 and Ar is constant at 0.18 and 1.08 Torr. The power supply is 1.9 kV. The deposition rate for this arrangement is 200 A min^{-1} at a target to substrate distance of 6 cm. Films deposited are about 4 μm in thick.

3.8 Electron Motion

Electron motion in a plasma is influenced by static electric and magnetic fields, while the ion mobility is only slightly reduced by the magnetic field.

3.9 Plasma Sheaths

The rate of transfer of electrons and ions from a plasma to an adjacent surface occurs at different rates because of the different masses. Hence a space charge region, in which one species is largely excluded, forms adjacent to such surfaces. This is called a sheath. The variation of potential between surface and plasma is largely confined to this layer. Except for conditions of very high current density at anodes, this space charge region will contain primarily particles of the low mobility species. The cathode dark space is positive.

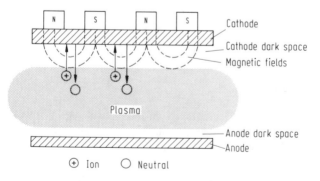

Fig. 3.9 DC planar magnetron sputtering (from Brodie, Muray 1982).

3.9.1 Ion Nitriding

Nitride surfaces can be built up on steel surfaces from a plasma, in which nitrogen ions are accelerated to the parts to be treated. Installations of considerable size can be employed in which a vacuum is maintained and the process gas is led into the apparatus through a regulating valve. Temperatures required for ion nitriding are lower than those for conventional gas nitriding. The gas employed is normally a mixture of hydrogen and nitrogen. The temperatures are in the range 350 to 580 °C. The time required is a function of case depth.

3.10 Magnetrons

Magnetic fields are used to deflect electron from striking the substrate. Various magnetically confined discharge configurations have evolved. A magnetron source is a high rate means for transferring material to a substrate and maintaining composition, e.g. coating single crystal of Si with Al-Si in integrated circuits. Figure 3.10 shows a coaxial magnetron configuration.

The magnetic field is axial, and the electrodes are in the form of connected cylinders. The electrons from the cathode surface are trapped by the crossed electric and magnetic fields in the dark space. They can escape only by losing energy in ionising collisions.

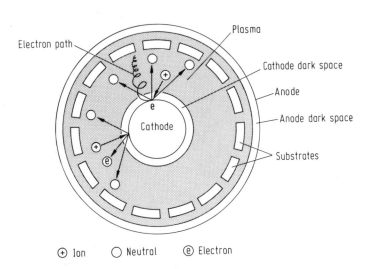

Fig. 3.10 Coaxial magnetron configuration (from Brodie, Muray 1982).

3.11 Coating of Plate Glass and Plastic Sheet by Planar Magnetrons

Plate glass is coated with films for both solar purposes and heat reflection using planar magnetrons up to 3 metres long. Input power is up to 80 KW on one target. Additionally, plastic sheet is coated for different purposes including dielectric interference.

3.12 Ion Implantation

Ion implantation was developed by the semi-conductor industry as a technique for introducing controlled levels of dopants into surfaces.

It has also been applied to ceramics where the implant affected area is $<$ 0.5 μm thick. It influences hardness and indentation fracture behaviour of ceramic materials. Hardening occurs by point defects which hinder dislocation motion in the surface, and by solid solution effects. Above a critical dose, amorphous material grows which may lead to softening.

A variety of ions can be used as implants.

An ion having sufficient energy (10–500 keV), striking a solid surface will have a probability approaching unity, of entering the structure. It will come to rest following a sequence of collisions with atoms. Figure 3.11 shows schematically a depth concentration profile of a low ion fluence implanted into a solid. A commercial ion implantation system is shown in Figure 3–12.

In this system, the dopant atoms are ionised, and accelerated to 30–35 keV by an electric field. A mass separating magnet eliminates unwarranted ion species.

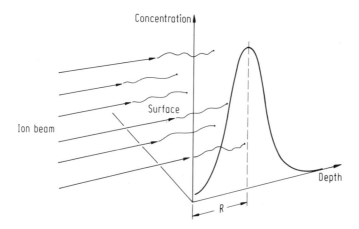

Fig. 3.11 Depth concentration profile in an ion beam implantation proces on a surface.

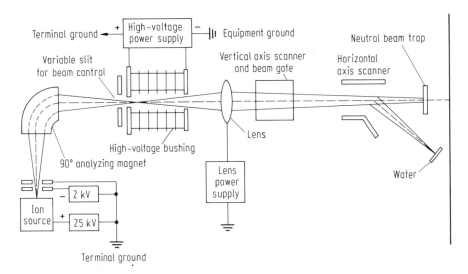

Fig. 3.12 Commercial ion implantation system.

3.13 Spraying of Metals and Oxides by Gas Flame, Arc and Plasma

3.13.1 Gas Flame and Arc Methods

Metals and oxides may be sprayed in the form of particles onto surfaces. Gas flame and arc techniques melt the metal from wire and the spray is formed by impact of a high velocity gas stream on the molten material. Figure 3.13 shows an arc melting apparatus. The arc is struck between the two wire electrodes and compressed air is injected to transfer particles to the work.

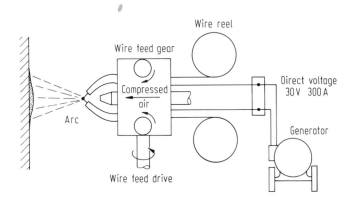

Fig. 3.13 Arc Melting equipment for spraying.

The metal surface must be prepared initially for a good bond to occur between the particles. Adhesion is normally higher than in flame sprayed coatings. The velocity of particles in arc spraying is lower than with flame spraying but the particle size is greater and the temperature is higher. Typical applications are zinc and aluminium.

3.13.2 Plasma Spraying

Conventional gas spraying, and also arc processes have a problem in oxidation of the particles. In the plasma processes, the technique is based on the injection of powder into a high temperature inert gas plasma. Since there remains also a problem of oxidation of droplets, low pressure techniques have been developed for plasma processes (LPPD-low pressure plasma deposition). Low pressure plasma deposition is performed in an evacuated chamber. Surfaces can be built up and thin-walled parts can be manufactured. The injected particles melt and resolidify on the substrate. The plasma jet operates at 10^4K and higher. Ceramics can be sprayed.

A typical plasma spray gun is shown in Figure 3.14. This is a transferred arc plasma flame torch. The arc is vortex stabilised by gases injected tangentially behind the cathode. The gases are selected for inertness or energy content and include Ar, Ar-He, Ar-H_2, N_2, N_2-H_2. Power levels may be 80–120 kW. The injected powder particles are introduced either in the nozzle throat or beyond the nozzle exit. The long throat increases heat transfer to the powder.

In LPPD, a low pressure inert gas atmosphere is provided in a water-cooled chamber. The plasma with entrained particles is sprayed into the chamber and onto the substrate. The gas flow rates are 100–250 ccmin^{-1}.

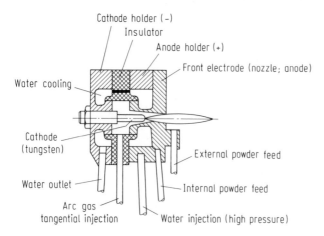

Fig. 3.14 Plasma spray gun.

Basically, these spray processes do not create molten layers since the particles solidify individually. The structure obtained is characteristic of a sprayed particle process. By comparison, arc, electron-beam, plasma melting and laser processs used in the manner described in Chapter 2 and in Chapter 3 are surface melting processes and the structures are typical of cast material solidifying from a molten layer. The spraying method solidifies particles with individually refined structures and the coating is obtained solidified particle by particle. Where measurements have been made of the strength of sprayed deposits, the figures show increases over conventional cast strength. This is related to fine grain size and structural effects noted for rapid solidification (see Chapter 1) including extended solid solubility of solute elements.

Plasma sprayed deposits can be employed as follows:

Protective coatings
Erosion and Wear Protection
Thermal barriers (also termed PTBC, Protective Thermal Barrier Coatings)
Build up of worn parts

Interesting application are in net shaped manufacture and in preparing ceramic linings.

3.14 Chemical Vapour Deposition CVD

Chemical vapour deposition processes depend on chemical reactions at the surface of a substrate. A coating is formed plus volatile products. A typical reaction forms Silicon films as follows

$$SiH_4 \rightarrow Si + 2H_2$$

Figure 3.15a shows a typical reactor vessel for CVD. The vessel can be hot walled or cold walled Figure 3.15b. The gases are led into the reactor and flow over the stacked pieces for coating. The reaction products are removed from the vessel. The processes can be operated at atmospheric or at reduced pressures. They are of much importance in electronics and in tool manufacture. Important new aspects of these processes also include diamond films and superconductors. Tool manufacture uses CVD processes to deposit TiC and TiN on tool surfaces. The reaction is controlled by the thermodynamic equilibrium which depends on vapour pressure and temperature.

Different CVD processes are distinguished by the temperature of the substrate:

High temperature CVD $\quad$ $TiCl_4 + N_2 + H_2 \rightarrow TiN + HCl$
$850\ °C < T < 1200\ °C$

Moderate temperature CVD $\quad$ $TiCl_4 + CH_3\ CN + H_2 \rightarrow Ti(C, N) + CH_4Cl$
$700\ °C < T < 850\ °C$

Low temperature CVD $\quad$ $TiCl_4 + N_2 + H_2 + Ar \rightarrow TiN + HCl + NH_3$

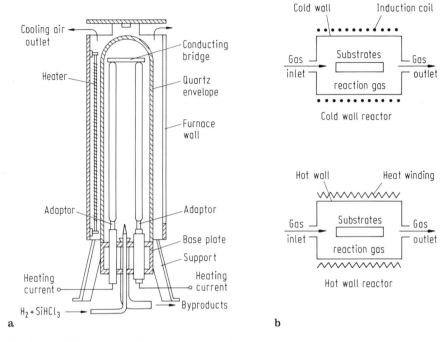

Fig. 3.15 (a) CVD reactor. (b) Cold wall and hot wall reactor vessels.

CVD processes can be compared with the other coating processes such as evaporation and sputtering. These processes are classified as PVD (physical vapour deposition). The CVD processes are performed at higher vapour pressures and higher temperatures than PVD. Single, double and triple layers can be formed in CVD while PVD is normally confined to single layers.

References

Apelian D, Szekely J. (1987). Plasma Processing and Synthesis of Materials MRS.

Arata Y (1986). Plasma, Electron and Laser Beam Technology ASM.

Ashby E, Easterling K (1984), Laser Transformation Hardening, Acta Met. 32 1935–1948.

ASME (1982). Explosive Welding and Forming ASME.

Birkhoff G, MacGongal DP, Pugh E, Taylor G. J. Appl. Phys. 19, 6:563.

Brodie I, Muray JJ, (1982). The Physics of Microfabrication, Plenum.

Bunshah R (1982), Deposition Technology for Films and Coatings. Noyes Publications.

Carter G. Colligon JS, Grant WA (1975), Applications of Ion Beams to Materials. Conf. Series No. 28, Inst. of Physics, London.

Duley WW (1976). CO2 Lasers. Effects and Applications. Academic Press.

Ganin E (1984). D.Sc. Thesis Dept. Mat. Eng. Technion, Haifa.

Kossowsky R, Sing H (1984). Surface Engineering AIME.

Ku TN, Rosenberg R (Eds) (1982). Preparation and Properties of Thin Films. Treatise on Mat. Sc. and Tech. Academic Press.

Lyakhovich LS, Voroshuin LG, Paurich GG, Scherbakov ED (1974), Multicomponent Diffusion Coatings (published in Russian). Published in English (1987) NBS. Washington, USA.

Spalding IJ (1987), How to Select a Suitable Laser. In: Proc. 4th Int. Conf. Lasers, IFS Ltd. Manufacturing Conferences.

Chapter 4

Powder Processes/Hard Metals/Ceramics/ Glass Ceramics

4.1 Introduction

Powder processes are used for the production of a broad variety of materials. These involve metals and oxides in forming techniques as powders which are pressed into shape and sintered. Sintering occurs by the transfer of matter into the voids between particles by solid state or by liquid processes. The present technology for powders produced by rapid solidification processes makes use of compaction followed by shape forming using plastic processes. For many materials e.g. tungsten, metal carbides, and ceramics, sintering must be employed in shape forming, because the high melting point, or the physical properties of the materials, do not presently allow any other procedure.

4.2 Powder Production Methods

Different methods of powder production are possible. Examples are:

1. Reduction of oxides
2. Atomisation and Rapid Solidification
3. Precipitation or co-precipitation
4. Growth from a Carbonyl Phase
5. Thermal decompostion of salts
6. Sol-Gel Methods
7. Growth from the Vapour Phase.

4.2.1 Reduction of Oxides

Different processes may be employed to reduce oxides and produce metal powders, a common procedure being the use of hydrogen, e.g. in the reduction of tungsten oxide. Gaseous reduction processes may be controlled to give particles of required geometry for subsequent pressing, the shape and size of particles depending on temperature, gas flow and initial particle size.

Typical reduced powders are W, Mo, Fe, and Co. Tungsten is reduced from WO_3 which is precipitated from tungstic acid solution, prepared by acid treatment of the tungsten ores Wolframite and Scheelite. The precipitated oxide undergoes purification treatment and is reduced by hydrogen at about 850 °C. Particle sizes of 1–3 μm are produced for applications such as lamp filaments and 20 μm for applications such as X-ray targets. Iron powders can be made by reducing pure iron ore or mill-scale, which is the oxide produced on steel in rolling processes.

4.2.2 Gas Atomisation and Rapid Solidification Methods

A liquid stream of metal, falling through an impinging gas stream is torn into ligaments and in turn into drops. These are collected as powder in a hopper at the base of the apparatus. The process was described in 1.20.3.

In some methods, the gas is driven upwards past the metal held in a container and a film of metal at the container periphery is atomised. The powder is collected from the gas stream.

The methods are common to rapid solidification technology and the powder has a rapidly solidified structure, (see Section 1.20). This technology is employed for preparing alloys of high strength and for alloys which are difficult to cast. The techniques may be used to make powder which is subsequently consolidated by extrusion or rolling.

4.2.3 Centrifugal Atomisation

A type of apparatus, suitable for breaking up the liquid by centrifugal force is shown in Figure 4.1. An electrode is arced against a metal cathode rotating as a disc and drops are thrown off along the conical surface to solidify as powder.

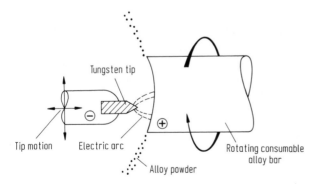

Fig. 4.1 Centrifugal atomisation using a tungsten cathode and conical surface by Nuclear Metals Inc. (from Metals Handbook 1984).

4.2.4 Sol-Gel Techniques

These techniques can be used to produce low density spheres of high surface area and high chemical activity. Such spheres can be sintered to theoretical densities at unusually low temperatures.

A sol is a suspension of colloidal particles formed by controlled precipitation from aqueous solutions. Gelling to a semi-rigid body is done by evaporating the water, or by changing the pH. In many cases, the sol is first broken into spherical droplets before gelling. This sol may also be dehydrated by passing the droplets through a column of dehydrating liquid.

For pH control, an ammonia donor, e.g. hexamethylene tetramine, is added to the sol. The droplets are passed through a hot organic fluid which causes the evolution of ammonia. Further drying in both cases gives low density spheres.

Gels may be formed by polymerisation of metal oxides. This is important in the formation of homogeneous ceramics and glasses at low temperature.

Aerogels are gels dried by having the system enter a liquid-solid phase area where the vapour phase is absent and the solvent is removed only as liquid. Aerogels have low densities and are used in insulation.

Sintering of gels can be made at low temperatures, much lower than those for normal sintering of oxides or for fusing of a glass. As an example, silica, which has a fusion temperature of 1750 °C can be prepared by a sol-gel process at 700 °C.

The sol-gel method is described as follows. The components are mixed to form the sol e.g. by controlled hydrolysis. The solution sets into a stiff gel. The gel particles interconnect to form a network containing an interstitial liquid phase in its meshes. An aging process sets in, which is one of Ostwald ripening and further solute deposits. The interstitial phase is next eliminated by drying and the soft jelly is progressively transformed into a porous solid. Very fine powders can be produced by milling. Alumina can be produced by a sol-gel process. As discussed, glasses have been made which normally require high fusion temperatures and which might crystallise as they cool down.

4.2.5 Powder from Carbonyl

One example of powder production from a carbonyl is that of the manufacture of pure iron powder by decomposition of iron carbonyl $Fe(CO)$. The carbonyl is decomposed at 150–400 °C and atmospheric pressure. Iron carbonyl is a liquid boiling at 103 °C. Spherical particles of iron are produced by mixing the carbonyl with an inert gas and allowing the decomposition to powder in a vessel without contact on hot surfaces.

4.3 Powder Consolidation

Figure 4.2 shows possible routes for the production of parts by powder metallurgy, and shows the conventional route compared with HIP (described in the following section) and hot compaction.

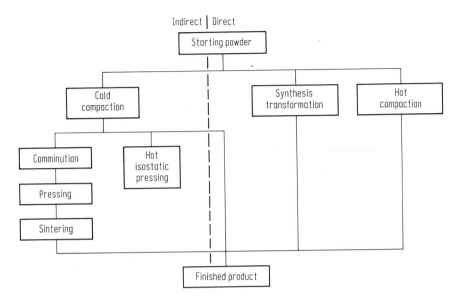

Fig. 4.2 Possible routes in powder metallurgy (from Metals Handbook 1984).

Figure 4.3 shows the conventional method in which powder is mixed with lubricant and a binder. The powder is consolidated in a mould under pressure and ejected as a green compact which is then sintered in a furnace. The sintering process is directed to the removal of porosity between particles, (the theory of sintering is given in Section 4.4). In the pressing or compaction stage, the density of the part is related to the distribution of stress in the mould, as well as to the shape and the size of particles. The sintering stage attempts to arrive at 100% density of the part and in effect, the final strength of the sintered material is related to its density.

In the following paragraph, injection moulding is described which is a technique employed to improve both productivity and green density.

4.3.1 Injection Moulding

This technique, consists of the following. The metal powder is mixed with a plastic medium, comprising an organic binder (methyl cellulose) dissolved in water. Additions of materials like glycerin/boric acid are incorporated to promote mould release. This mixture is injected under pressure into a closed pre-heated die. The compact becomes sufficiently self-supporting to hold its moulded shape during ejection from the die, and is dried and sintered.

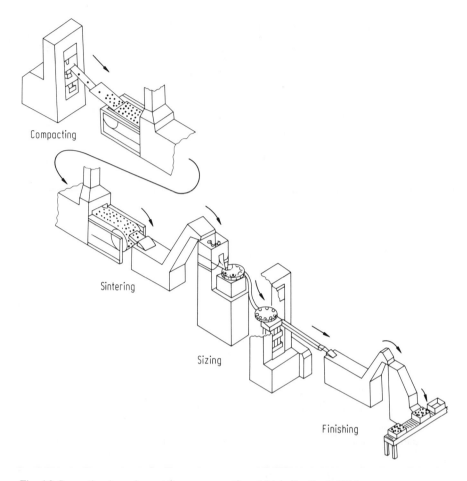

Fig. 4.3 Conventional powder metallurgy process (from Metals Handbook 1984).

4.3.2 Dynamic Compaction

Dynamic compaction is used for the bonding and densification of otherwise difficult to form particulate materials, e.g. carbides, nitrides etc. Dynamic methods include the following:

Compressed air compaction
Gas gun compaction
Magnetic dynamic compaction
Hydrodynamic compaction
Explosives

Examples of materials which are compacted by these techniques are:

BN   Boron Nitride
AlN   Aluminium nitride

B_4C Boron carbide
SiC   Silicon carbide
Si_3N_4 Silicon Nitride

The range of compaction times and pressures are 1 to 100 sec and 10 GPa. The intense pressure pulse during compaction is accompanied by changes which are different from the usual kinetic processes for densification. The shock-wave phenomena are expressed in the Rankin-Hugoniot relationship.

Undesirable gradients of density with accompanying gradients of properties occur which may lead to cracking.

4.3.3 Hot Isostatic Pressing (HIP)

HIP techniques are applied to consolidate materials such as tool steels, superalloys, Ti alloys, ceramics etc. The process of pressing is performed hot and the temperature contribution is important for materials where long sintering times are related to low green densities e.g. ceramics. Figure 4.4 shows an autoclave in which a powder in a metal container is heated while subject to an isostatic gas pressure. The powder is sealed in a flexible container and vacuum degassed. The container is a metal can for simple shapes while for complex forms a ceramic glass mould is used.

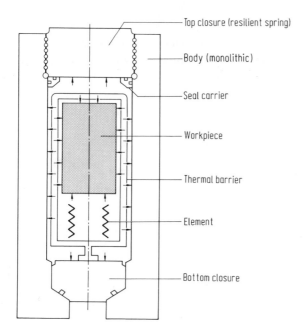

Fig. 4.4 HIP process.

In the normal HIP process, the powder is preformed by a variety of techniques including cold isostatic pressing, die pressing, vibratory filling etc. After preforming, the body is hermetically sealed in the container which is impervious to the gas at high temperature and pressure.

A number of mechanisms contribute to densification of the compact, and these include the radial symmetric closure of pores, plastic deformation, dislocation creep, diffusional flow (Nabarro-Herring Creep) and Coble flow. In addition, densification may be influenced by superplasticity.

For super alloys, the process can be conducted above the yield stress when plastic deformation will cause appreciable densification. Densification by plastic flow will still be less than the desired value so that creep and the deformation modes given above will be required. Densification maps have been published and show the relative density changes as a function of P/σ_y where P = pressure (Figure 4.5). The map shown is for a gas atomised tool steel powder. The initial densification process is one of yielding. As the process proceeds with time, the maximum density is achieved by a combination of time-dependent power law creep and diffusion.

4.4 Sintering

There are two main processes for sintering:

1. Solid State Sintering
2. Liquid State Sintering

The mechanisms of the two processes are different and in the present section, the solid state process is described. In this, the driving force for diffusion is theoret-

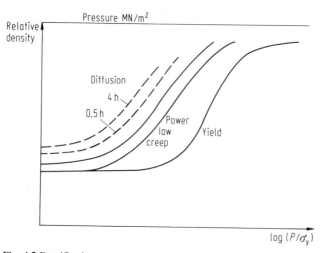

Fig. 4.5 Densification map.

ically related to the reduction of surface energy. Diffusion occurs through different paths on the surface of the particles, along grain boundaries or through the lattice. Sintering is normally performed without pressure but the densification process is assisted by pressure in processes such as HIP (described in previous section) hot forging, and hot extrusion.

4.4.1 Solid State Model of Sintering

A model of the sintering process is shown in Figure 4.6. Two spheres pressed in contact form a neck which then changes its geometry by the flow of solid in a diffusion process (mass transport), promoted by high temperature.

In the model, Figure 4.6, the neck radius is ρ. Applying an analysis of concentrations of solute or vacancies in equilibrium with surfaces of varying curvature, using the Gibbs-Thomson relationship, it may be shown that at the surface of radius ρ, there is an excess of vacancies over the concentration of vacancies present at the original sphere radius. The vacancy concentration gradient provides a flow of vacancies away from the neck in exchange with a diffusional flow of atoms to the neck. The driving force for the geometrical change is the reduction of surface energy and this reduction occurs with neck growth. The next stage of the sintering process is the removal of pores, and the kinetics in this stage is related to the pore-solid interfacial area, with transport mechanisms by volume or grain boundary diffusion.

4.4.2 Sintering Maps

Sintering maps which describe the solid state process were suggested by M. Ashby, Figure 4.7. The diagrams are for the different sintering mechanisms, and lines of equal contribution from each mechanisms are drawn on axes representing neck size and reduced temperature, (T/T_m) where T_m = melting temperature. Different areas of the diagram show the mechanisms which dominate within each region.

4.4.3 Liquid Phase Sintering

Liquid phase sintering is also called activated sintering. It is an important sintering process because of the rapid kinetics and the ability to produce useful forms from difficult starting materials, e.g. in the manufacture of carbide cutting tools.

The liquid phase can be obtained in different ways. Usually it is a part of the starting system and originates from the composition of the material, which has a liquid phase present at the sintering temperature. Alternatively material is added separately as a solid powder which becomes liquid at the sintering temperature e.g. metallic binders in carbide manufacture. Examples are Fe, Ni, Co.

The liquid in this type of process is the active medium for transporting matter as well as having other functions. This is an important difference between liquid phase and solid phase sintering. Applications of liquid phase sintering are found in

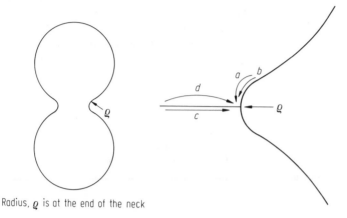

Radius, ϱ is at the end of the neck

Fig. 4.6 Model of solid state sintering process showing diffusional flow of atoms to the neck.

metals, ceramics, firebrick, porcelain, ferro-electric capacitors, ferrite magnets. The driving force for sintering in these systems is the sum of the decrease of interfacial free energy, the overall volume free energy of the system and the stored energy. Transformations during the sintering process may contribute to a decrease of the free energy while the decrease of stored energy results from dislocation movement and from grain growth.

An overall view of the mechanisms in liquid phase sintering is given in Figure 4.8. The mixed powders are first shown in Figure 4.8(a), followed by the formation and the spreading of liquid and its effect on particle re-arrangement. In Figure 4.8(b) the different phenomena of precipitation, diffusion, grain growth and shape accommodation are shown. In 4.8(c) solid state processes are shown. The densification mechanism is described as occurring in three overlapping stages as follows.

4.4.3.1 Formation of liquid

The liquid exerts capillary force on the solid particles which promotes densification. The compact behaves as a viscous solid, and the particles undergo rearrangement. Factors which affect densification are the amount of liquid, the size of the particles and the solubility of solid in liquid. Particle rearrangement improve with fineness.

4.4.3.2 Particle coarsening

In the second stage, solution, reprecipitation, and micro-structural coarsening occur. Small grains dissolve and larger grains grow. The re-arrangement of particles slows and solubility effects and diffusivity dominate.

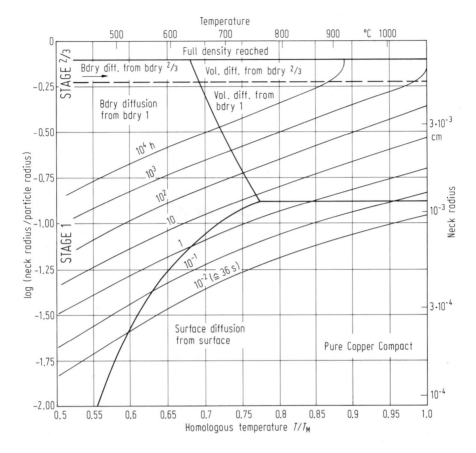

Fig. 4.7 Sintering map (from Ashby 1974).

4.4.3.3 Solid state sintering

In the final stage, the densification process occurs by a mechanism of solid state sintering allowed by contacts between grains in the solid state.

4.5 Hard Materials

This is a generic term for materials used effectively in metal cutting. Hard materials include metal carbides, oxides, nitrides, diamond. An important part of the hard materials industry is related to WC and to TiC. For the metal cutting industry in general, cutting tools fall into the categories of High Speed Steel, WC, TiC, Cermets, and Ceramics. Hard materials generally are produced by different processes with powders and new processes by surface technology are being introduced. High

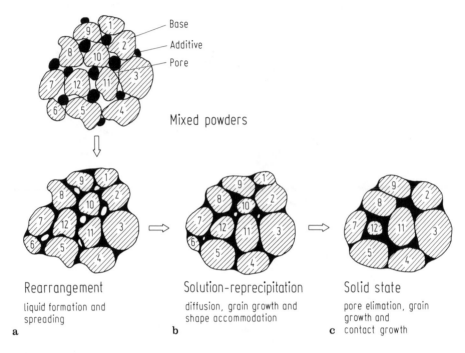

Fig. 4.8 Liquid phase sintering stages (from German 1985). (a) Mixed powders followed by first stage of sintering, which is formation of liquid. (b) Particle coarsening. (c) Solid state sintering.

speed steel parts, formerly made from mechanically worked profiles, are now also manufactured by a powder process.

4.6 Ceramic Materials

A ceramic material is one which is capable of withstanding high temperatures and it is also a generic term for a group of materials having high hardness (see previous section). Industrial use is made of a variety of types of ceramic. In the present section, some of the powder processes employed in ceramic fabrication are described. Other processes, e.g. those involving fibres, are given in other chapters. Some typical ceramic materials are given as follows:

Oxides	Al_2O_3 SrO SiO_2
Carbides	WC TiC SiC
Diamond	
Graphite	
Nitrides	ZrN Si_3N_4 BN
Borides	Fe_2B TiB_2
Sialon	See Section 4.6.4.

4.6.1 Manufacture of Sintered Carbides

Sintered carbides for cutting tools are principally tungsten carbide bonded with cobalt. The other carbides which may be present are Ti, Ta, Nb, Cr, V, Mo and Hf. In addition to Co, the metals Ni and Mo may be added for liquid phase sintering.

The method of manufacturing WC is as follows:

Tungsten powder is first prepared from the ore and WC is prepared by carburising the tungsten powder with carbon black at temperatures between 1400 °C and 1500 °C. It is also possible to manufacture WC by mechanical alloying in a process which involve W powder and carbon.

After the WC manufacturing process, the metal carbide powder or powders are mixed with finely divided metallic binder principally Co and ball milled. 0.5–2% lubricant is added for the stage of compaction. The lubricants employed use paraffin wax in carbon tetrachloride or xylol, or beeswax and camphor in petroleum ether. The carrier is afterward removed by distillation under reduced pressure at which stage the carbide particles should have coatings of the metallic binder followed by the lubricant. The powder may be pelletized by milling to agglomerate pores and facilitate die filling.

The powders are pressed in hydraulic machines at 50 to 150 MNm^{-2}. Some of the pressings may weigh up to several kgs. Cold isostatic pressing may also be employed, giving more uniform density in the compact material.

The pressed compacts are then heated in a protective atmosphere, and pre-sintered at 700–750 °C. The green compacts can be cut or stamped. Sintering is performed at 1350 °C to 1655 °C in hydrogen. In this operation 45–60% by volume of porosity is eliminated.

In using pressures up to 150 MNm^{-2} big advantages in quality of product are achieved but the process is expensive. Sintering may also be made by conventional means and HIP may then be used in a second operation.

4.6.2 Sintering of Ceramics/General Scheme

A general scheme of sintering for ceramic materials might include the following:

1. Mix powders.
2. Cold form or warm form the powders prior to sintering. Use isostatic pressing or uniaxial pressing (for large numbers). Press with or without binders and lubricants.
3. Extrusion, warm or cold, with addition of plasticisers, to produce continuous sections of pre-formed material.
4. Injection moulding, in particular for intricate shapes.
5. Slip casting in aqueous media, (See Chapter 1).
6. Machining of pre-forms in green state followed by sintering.
7. Sinter.

4.6.3 Oxide Purity of Starting Materials/Uniformity, Morphology, Grain Size/Sintering Aids

Different problems are associated with oxide sintering, some of which stem from the different behaviour of diffusing species in these systems when compared with metals. The strong covalent bonding means that diffusion is a slow process and heating to temperatures where diffusion becomes appreciable may reach the range where decompostion occurs. Grain growth at these temperatures can lead to lowering of strength of the product.

Uniformity of shape, the details of morphology and the grain size are important factors to consider in packing. Pore space needs to be a minimum. Non-uniformity of the powder leads to gaps in the continuity of solid and resultant holes in the product. Sintering aids are added to form a liquid phase for densification.

4.6.4 Sintering of β' SIALON

Sialon has been studied having the general formula $Si_{6-z}Al_z O_z N_{8-z}$ with $z = 0.5$. β'-Sialon is isostructural with β-Si_3N_4 and has the zinc blend structure. It is built up of 1, 2 and 3 dimensional arrangements of (Si, Al) (ON)$_4$ tetrahedra. It is prepared by mixing Si_3N_4, AlN and Al_2O_3 powders to give the stoichiometric composition. Sintering aids are added, e.g. 9 wt% Y_2O_3. The mixing is performed in an Al_2O_3 ball mill in hexane. If uniaxial pressing of the compact is made, the pressure is 29.4 MPa. For isostatic pressing, the pressure is 196 MPa. The compacts are placed in carbon crucibles which act as susceptors for induction heating. Sintering is for 1 hr at 1700 °C in nitrogen.

4.6.5 Sintering of β Si$_3$N$_4$

Si_3N_4 shows a unique combination of excellent high temperature properties, resistance to oxidation and thermal shock. However, it is difficult to sinter because the volume diffusivity is small as compared with that of the surface diffusivity.

Si_3N_4 may be sintered at 2100 °C under high N_2 pressure but suffers excessive grain growth. It is therefore hot pressed in graphite dies at about 30 MPa and 1800 °C. Metal oxides, when added for assisting densification by forming a liquid phase, form glasses at the grain boundaries. The following materials have been added as densifiers to assist glass formation: Mg_3N_2, AlN, CeN, ZrN, Mg_2S.

4.6.6 Cutting with β' Sialon

Table 4.1 shows a comparison of cutting of metals using β' Sialon, WC and Al_2O_3.

Table 4.1 Cutting performance of Co-bonded WC. Al_2O_3, and a β' — Sialon

		Cast iron	Hardened steel EN31	INCOLOY 901
WC	Cutting speed, m/min	250	5	20
	Depth of cut, mm	6.5		
	Feed rate, mm/rev	0.50		
Al_2O_3	Cutting speed, m/min	600	Impossible	300
	Depth of cut, m	6.5	to cut	No second
	Feed rate, mm/rev	0.25		entry
β' — Sialon	Cutting speed, m/min	1100	120	300
	Depth of cut, mm	10.0	0.5	2.0
	Feed rate, mm/rev	0.50	0.25	0.25

4.6.7 Glass Ceramics

Glass ceramics are described here. While not a powder process in the conventional sense for ceramic materials, this type of material and the method of manufacture may be included with processes related to ceramics. A glass part is first made and a crystalline solid obtained by nucleation.

Glass ceramics have the unique property of a non-porous microstructure. The raw materials, which are mainly those for conventional glass systems are oxides of the common elements Si, Al, B, Na, Ca etc., plus non-oxides e.g. fluorides, chlorides, sulphides etc. The different compositions are mixed in batches. They may be compacted into briquettes using a suitable bonding agent to minimise segregation and melting losses. After melting, gas bubbles are removed from the system using arsenic and antimony pentoxides. These release oxygen. Gas in the melt enters the oxygen bubbles, which makes them grow and brings them to the surface.

Forming of parts is carried out by spinning, pressing, blowing, rolling or casting. These operations are performed in the appropriate viscosity ranges, and Figure 4.9 shows viscosity as a function of temperature, and the different ranges for processing. Casting is performed in a low viscosity range.

Nucleation of solid in the glass is heterogeneous and is performed on heating, the nuclei being complex oxides. The heat treatment conditions of time at any temperature are critical. The heating rates are similarly important and play a critical role in the nucleation of the desired phases and the development of the ensuing microstructure. The glass is heated rapidly to the temperature where nucleation takes place, held for a sufficient time to allow for the kinetics of the nucleation process in glass and then raised in temperature. The time at the new temperature is calculated to allow the desired microstructure to be achieved and an overall treatment cycle of a few hours to a few days may be required.

The majority of these materials are aluminosilicate glasses nucleated with TiO_2 or ZrO. The common aluminosilicate phases are β-spodumene and β-quartz solid solutions. They have thermal stability, thermal shock resistance, and chemical durability.

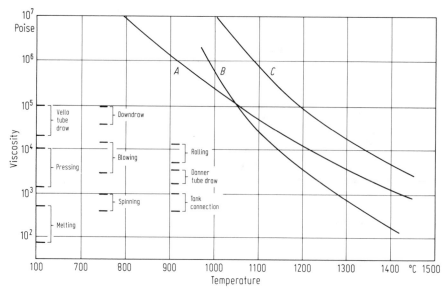

Fig. 4.9 Viscosity diagram for glass ceramics. Curves are for three different glasses (from Beall, Duke in Uhlmann, Kreidl 1983).

References

Ashby M (1974). Sintering Diagrams. Acta Met 22:275–289.

Alford MN. Birchall JD, Kendall K, (1986). Overview. Engineering Ceramics. The Process Problem. Mat. Sci. Technology 2:329.

Almond EA, Brookes CA, Warren A (1984). Science of Hard Metals. Inst of Physics, Adam Hilger Ltd.

Beall GH, Duke DA (1983) in Glass, Science and Technology (1) Uhlmann DR, Kreidl NJ (Eds) Academic Press.

Brinker CJ, Clark DE, Ulrich DR (Eds) (1984). Better Ceramics Through Chemistry. MRS.

Brookes KJA (1982). World Directory and Handbook of Hardmetals. 3rd Ed. Engineer's Digest and International Carbide Data.

Chin GR (Ed) (1982). Advances in Powder Technology ASM.

Das, Kear, Adams. Powder Met Int 10(3).

Doremus. Treatise on Materials Science and Technology Ceramic Fabrication Process 9:

German RM (1985) Liquid Phase Sintering. Plenum Press.

Grant NJ (1982). Powder by Rapid Solidification. In Advances in Powder Technology ASM.

Germans MEA (1973). Sol–Gel Process. Powder Met Int. 5:137.

Hench LL. Ulrich DR (1984). Ultrastructure Processing of Ceramics, Glasses and Composites. John Wiley.

Jack KH (1986) Sialon Hardmetal Materials in Science of Hard Materials. Almond EA, Brookes CA, Warren R (Eds) Inst. of Physics No. 75 Adam Hilger Ltd. Bristol.

Kingery WD, Bowen HK, Uhlmann DR (1976). Introduction to Ceramics. 2nd Ed. J. Wiley, Interscience NY.

Kingery WD (1959) Densification During Sintering in the Presence of a Liquid Phase 1. Theory, J. Appl. Phys. 30:301–306.

Klar E (Ed.) 1983) Powder Metallurgy. Applications, Advantages, Limitations. ASM.

Kolar D, Pejovnik S, Ristic MM (Eds) 1982. Sintering Theory and Practice. Elsevier.

Lawley A (1982) Gas Gun Compaction. in German RM, Lay KW (Eds), Processing of Metals and Ceramics AIME.

Mitomo M, Nagata S, Tsutsumi M (1983). Proc. First Int. Symposium Ceramic Compounds for Engines, Elsevier.

Tver DF, Bolz RW (1984). Encyclopaedia Dictionary of Industrial Technology, Materials, Processes and Equipment. Chapman and Hall.

Uhlmann DR, Kreidl NJ (1983). Glass, Science and Technology, Academic Press.

Chapter 5

Fibre Processes. Composites

5.1 Introduction

High strength is associated with materials having a covalent bond, e.g. Diamond. However, these materials have small values of fracture toughness and are susceptible to failure by rapid crack propagation.

The high strength of these materials is utilised in structural applications as composites, one type of which is of fibres of the high strength material dispersed in a matrix. The fibres need to be manufactured defect free or with a minimal defect content. They are dispersed in matrices which may be polymer, metallic, ceramic or glass, and the strength can be related to the volume fraction of the fibre phase. For a simple geometrical arrangement of fraction X_f of fibres, with strength σ_f, in a matrix of fraction X_m with strength σ_m, the tensile strength of the compact in the fibre direction is

$$\sigma_c = X_m \sigma_m + X_f \sigma_f \tag{5.1}$$

In this chapter, the different fibres and some of the different technologies available for their incorporation into composites, will be described. The manufacture of composite materials having a dispersed particulate phase will also be discussed.

5.2 Types of Fibre

Processes of manufacture have now been developed to manufacture a large number of materials in fibre form. A selected listing of fibres is given in Table 5.1.

5.3 Types of Composite

The types of fibre composite can be generally classified by the type of matrix and whether the fibre is continuous or chopped, Table 5.2.

The type of composite in which the fibre is replaced by particles, e.g. SiC is called "Particulate composite".

A brief listing of manufacturing methods for composites is given in Table 5.3.

Figure 5.1 gives a schematic representation of types of continuous fibre reinforced materials. Table 5.4 shows polymer matrix materials and some of the fibres with which they are used.

Table 5.1 Available fibre materials.

Type of fibre	Form available. commercial title
Boron	on Tungsten
Boron	on Carbon
Borsic	Boron Fibre with SiC coating
Graphite	Celanese GY 70
	Union Carbide. Pitch Base. (various moduli)
PAN graphite	Polyacrylonitrile, Union Carbide, PAN 300
	Union Carbide T50 Rayon; Hercules HM, HTS,
	AS-1, −3, −4
Alumina	Dupont FB
	Tyco
	ICI
Titanium Oxide	3M
Boron Nitride	
Boron Carbide	
Silicon Carbide	on Tungsten; on Carbon
	Spun from Melt Whiskers.
Metal Fibres	
Kevlar	

Table 5.2 Types of composite.

Matrix	Fibre type
Metal	Continuous or Chopped
Ceramic	Continuous or Chopped
Polymer	Continuous or Chopped
Glass or Glass Ceramic	Continuous or Chopped

Table 5.3 Manufacturing methods for composites.

Matrix	Method
Metal	Squeeze casting; Compocasting; Directional Solidification of Eutectic; Osprey Process. Thixotropic casting
Polymer	Pultrusion, Filament Winding.
Ceramic	Slurry mixing; Plasma Spraying

5.4 Fibre Manufacturing Processes

5.4.1 Carbon Fibres

Carbon fibres are a generic name for materials made by pyrolysis of polymer fibres in which the final heating operation serves to order the carbon atoms into hexagonal networks. The graphite lattice may not be strictly achieved and the layers might be termed turbostratic. The carbon atoms form the hexagonal network typical of graphite, aligned in the fibre direction.

A pyrolitic process is one involving heating and decomposition. The process

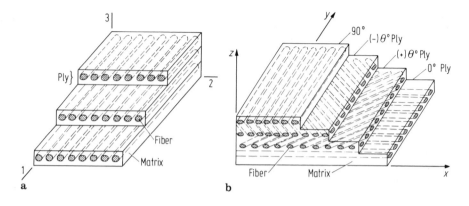

Fig. 5.1 Schematic representation of types of continuous fibre reinforced materials (from Lee, Mykkanen 1987).

Table 5.4 Polymer matrix materials.

Matrix Mateirals	Fibre
Epoxy, Phenol, Polyamide	Graphite, Kevlar, Boron, Glass

With acknowledgement to Willson and Dower, p. 752 Metals and Materials (1988), Vol 4.

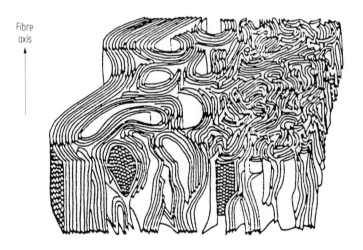

Fig. 5.2 Alignment of hexagonal planes in graphite fibre (from Hull 1981).

begins with a polymer which is first drawn. For PAN fibres, the starting material is Polyacrylonitrile, a polymer which closely resembles polyethylene. It has a nitrile group (CN) which replaces alternate hydrogen side groups of polyethylene.

The fibre is stretched to align the molecular chains along the fibre axis and then heated, first in an oxidising atmosphere and then in a reducing atmosphere where graphitising reaches a final stage.

The polymer active groups first interact to produce a ladder polymer consisting of a row of 6-membered rings. Further cross-linking occurs between the ladder molecules. In the reducing stage, and after all elements other than carbon have been removed, a structure of the hexagonal rings of carbon as in the graphite lattice eventually forms the basis of turbostratic graphite. The final structure is shown in Figure 5.2 consisting of oriented fibrils.

5.4.2 Silicon Carbide Fibres by Pyrolysis

Silicon carbide fibres by pyrolysis are not strictly stoichiometric in composition and have moduli lower than those of SiC formed by CVD. However they can be made by technological processes not strictly differing from carbon fibre manufacture and they are important fibres in reinforced composites.

The starting materials are organosilicon polymers which have a Si-C bond. An example is polysilmethylene.

Polycarbosilanes are obtained by using polydimethylsilane. This contains a few percent of a phenol group in place of the methyl group. They are capable of forming precursor fibres for SiC by melt spinning (see 1.17).

Pyrolysis and polymerisation of polydimethylsilane takes place under nitrogen gas flow at normal pressure. The fibres have O and H in the structure.

5.4.3 Glass Fibres

Different methods are employed for glass fibres. Figure 5.3.

This shows a furnace used for drawing fibres. The furnace is made of Pt-Rh and heating is by an electric current passing through the glass. Several hundred small nozzles are located in the furnace base and the fibres are drawn from these. On exit, the fibres are coated with an organic material and then woven into strands.

Typical compositions of glass used in fibres would range from 52–64% SiO_2, up to 25% Al_2O_3, Fe_2O_3, 17% CaO, 3–4% MgO, Na_2O, 9.6% K_2O, 10.6 Ba_2O_3%.

5.4.4 Kevlar Fibres

Kevlar is an aromatic polyamide fibre of high modulus, and fibres are made by processes which involve stretching and alignment of the polymer chains parallel to the fibril axis. The Kevlar structure is shown in Figure 5.4. It is composed of chains of aromatic carbon rings linked by -CO-NH-groups. After stretching, drying, and heating in nitrogen at temperatures upto 550 °C, the molecules form planar sheets bonded by H. The sheets are arranged in a random system of axially pleated lamellae. Figure 5.5.

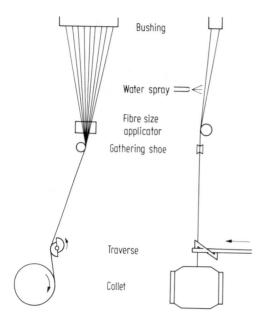

Fig. 5.3 Glass fibre production (from Loewenstein).

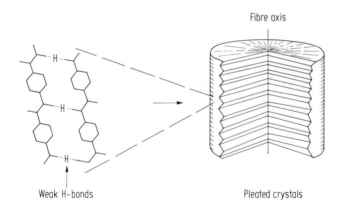

Fig. 5.4 The Kevlar structure.

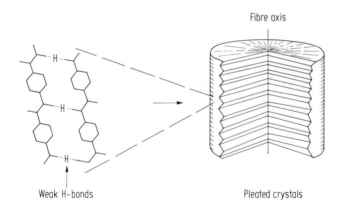

Fig. 5.5 Axially pleated lamellae of Kevlar.

5.5 Polymer Matrix Materials

5.5.1 Epoxy

The epoxy group is shown in Figure 5.6. Epoxy has good adhesion properties with fibres and low shrinkage in curing. There is an absence of voids. Adhesion between fibre and matrix is a problem in all fibre reinforced composite manufacture involving polymers, metals, glass, and ceramics. Under load, or in a chemical environment, separation may occur at the interface between these materials and the fiber. Of the polymer matrix materials, epoxy has favourable properties.

5.5.2 Other Matrix Materials

Thermo-setting matrices have significant application, particularly in aircraft, and thermoplastic matrices are also used. Complex shapes can be made by a simple moulding and curing process using a pre-preg or pre-cursor. This is an arrangement of fibres e.g. in bundles or sheets, impregnated with a reactive resin. Heating of the pre-preg melts the resin and the material can then be moulded to the required shape. The quality of the precursor can be optimised by filtering out strength limiting defects, and by optimising the processing conditions. In thermoplastic matrices, use is made of polyethersulphone (PES) and PEER (polyether etherketone).

5.6 Pultrusion

Pultrusion is one of the methods for composite manufacture with a polymer matrix. It is a continuous moulding process in which fibres are pulled through a thermosetting resin, normally a polyester, Figure 5.7. Consolidation and curing take place in a die. The composite produced by the process is then cut to the desired length. The die temperature is 150–180 °C. The pulling speed varies with the resin and is about 0.1 m.min^{-1} for epoxy systems. The reinforcement can be a uni-directional fibre, a bi-directional fabric, a random net etc. The resins can be polyester, vinyl ester, epoxy and others. Thermoplastics can be post-formed after pultrusion.

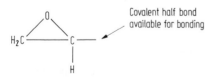

Fig. 5.6 Epoxy group.

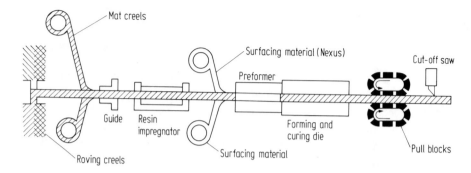

Fig. 5.7 Pultrusion (from Liskey, 1989).

5.7 Cast Metal Composites

5.7.1 Compocasting of Particulates

Metal matrix composites can be manufactured by various casting processes. The composite can be particulate, i.e., a dispersion of particles in the matrix, or a fibre composite with continuous or short fibres. The composites of Al with dispersed phase Al_2O_3, SiC, graphite etc., can be made by melting the metal, mixing the dispersed phase by a stirring operation and pouring. The technique involving stirring of the phase to be dispersed in a semi-solid liquid is called compocasting. Manufacture can also be made by the Osprey process (Chapter 1), by injecting solid particles of the phase to be dispersed into the stream of atomised liquid.

Different problems arise in the process, e.g. carbon fibres are not ordinarily wetted by Aluminium, and Aluminium and Carbon react to form Aluminium Carbide above approximately 500 °C. The problem of wetting can be solved by coating the fibres. Changing the alloy compostiion e.g. using Al-Si can avoid the problem of a reaction between Aluminium and Carbon.

5.7.2 Liquid Metal Infiltration Processes

For composites of fibers, the problem becomes one of infiltration of metal into the space between the fibres. This process can be performed by pressure, e.g. by squeeze casting.

An example of a cast metal matrix composite is the manufacture of Al pistons with a ceramic fibre insert to reinforce the piston ring groove. This has an important influence on piston wear without influencing thermal conductivity of the piston.

The minimum pressure in squeeze casting processes has been calculated from the following.

$$P = \frac{AV_f(\sigma_{SL} - \sigma_{SA})}{d_f(1 - V_f)}$$

σ_{SL} = surface energy between particle or fibre and the liquid.

σ_{SA} = surface energy between particle or fibre and the atmosphere.

d_f = particle or fibre diameter.

V_f = volume fraction of the reinforcing phase.

A = a factor which has the value 4 for cylindrical particles and 6 for spheres.

The equation is correct only if irreversible chemical reactions at the interface do not take place. These however occur and influence surface energy. The procedures adopted therefore may require pre-treatment of the reinforcement, modification of alloy composition or coating of the reinforcement. CVD coatings are widely employed.

In addition to capillary forces calculated in the above equation, friction forces due to metal viscosity require additional pressure.

One of the problems to be overcome is shrinkage porosity and risering of castings is important.

5.8 Ceramic Matrix Composites (CMC)

5.8.1 Fibre Reinforcement

Fibre reinforcement of ceramics, i.e. having a ceramic matrix toughened by fibres is a possible mode of increasing the low fracture toughness of ceramics. Such materials may also be capable of operating at elevated temperatures.

Examples are given as follows:

1. Carbon, or SiC, or Alumina fibre toughened glass and glass ceramics.
2. SiC fibre toughened Alumina matrix or Si_3N_4 matrix materials.

On failure of the matrix, the load is taken up by the fibre, and bonding between fibre and matrix is important. Deflection of a crack by a fibre is also an important toughening mechanism to be considered.

Zirconia matrix materials are toughened by a phase transformation of the martensitic type, which involves a change of structure ahead of the crack.

One method described for CMC manufacture is slurry mixing and infiltration. A preform of fibre material is first made and impregnated with the matrix material by passing it through a slurry, which is made up of the ceramic powder in water or alcohol, with an organic binder. After drying and cutting to shape, the organic binder is burnt out, and densification is obtained by sintering.

5.8.2 Carbon-Carbon Composites

These consist of carbon fibres in a carbonaceous matrix, which can have different structures including glassy carbon. This is a glass having carbon to oxygen bonds.

The matrix can be made by gas or liquid phase techniques. During carbonisation, pressure is applied by HIP.

5.9 Glass Ceramic Composites

5.9.1 Glass Ceramics

Glass ceramics may be reinforced by fibres in the casting process.In the composite, reinforcement is obtained by, graphite, silicon carbide, or aluminium oxide.

References

Ahmad I, Noton BR (1980). Advanced Fibres and Composites for Elevated Temperatures. Met. Soc. AIME.

Chawla KK (1981). Composite Materials Science and Engineering. Springer-Verlag.

Delmonte J (1981). Technology of Carbon and Graphite Fibre Composites. Van Nostrand, Reinhold.

Gray G, Savage GM (1989). Advanced Thermoplastic Composite Materials. Metals and Materials 5(9):513.

Harrigan WC, Strife J, Dhingra AK (1985). Fifth Int. Conf. on Composite Materials [CCM-V Met Soc Inc.

Hench LL, Ulrich DR (1984). Ultrastructure Processing of Ceramics, Glasses, Composites. John Wiley.

Hlavac J (1983). The Techonology of Glass and Ceramics. An Introduction. Glass Science and Technology (4) Elsevier.

Hull D (1981). An Introduction to Composite Materials. Cambridge Univ. Press.

Kawate K, Aksata T (Eds) 1982. Japan-US Conf. Composite Materials. Applied Science Publishers.

Kelly A, Rabotnov YN, Eds (1985). Handbook of Composites. Vol. 1. Strong Fibres. N. Holland.

Lee JA, Mykkanen DL (1987). Metal and Polymer Matrix Composites. Noyes Data Corporation.

Liedtke MW, Toad WH (Eds) (1985). Advanced Composites ASM.

Liskey AK (1989). Pultrusion on a Fast Track. Adv. Mat and Proc. 135 (2):31–35. ASM

Loewenstein KL (1973). The Manufacturing Technology of Continuous Glass Fibres. Elsevier.

Meyer RM (1985). Handbook of Pultrusion Technology. Chapman and Hall N.Y.

Pajget O, Reichstadter B (1979). Processing of Polyester Fibres. Elsevier.

Petzow G (1984) Nature and Structure of Metal-Ceramic Interfaces. in Upadhyaya (Ed.) Sintered Metal Ceramic Composites. Elsevier.

Schwartz MM (Ed) (1985). Fabrication of Composite Materials. Interfaces. in Upadhyaya (Ed.) Sintered Metal Ceramic Composites. Elsevier.

Schwartz MM (Ed) (1985). Fabrication of Composite Materials. Source Book ASM.

Vinson JR, Sierakowski RL (1986). The Behaviour of Structures Composed of Composite Materials. Martinus Nijhoff.

Watt W, Perov BV (Eds) (1985). Strong Fibres. N. Holland.

Willson MC, Dower RJ (1988). Metals and Materials 4:752–756.

Chapter 6

Shape Forming

6.1 Plasticity of Metals/Physical Definitions

Plasticity of metals is concerned with ductile behaviour when the material is stressed beyond the elastic range. This has important applications in Engineering related to metal forming.

Some definitions and relationships are given as follows:

Engineering stress $\sigma = F/A_o$
Engineering strain $\varepsilon = (l_1 - l_0)/l_0$
True stress $\sigma_0 = F/A$

True strain (logarithmic strain) $\varepsilon_0 = \int_{l_0}^{l_1} \frac{dl}{l} = \ln \frac{l_1}{l_0}$

A_0 = Initial cross section
A = Actual cross section
F = Applied load
l = Length
l_0 = Initial length
l_1 = Final length

A diagram for steel showing Engineering Stress and Engineering Strain shows a yield point and a maximum stress at which necking occurs. A diagram for steel showing true stress and strain is continuous.

In plastic deformation processes, it is convenient to work with true strains. These are additive. Strains in materials resulting from different parts of the process, or from different processes may then be conveniently added.

$$\varepsilon_3 = \varepsilon_2 + \varepsilon_1$$

6.2 Plasticity Theory/Yield Criteria

6.2.1 Tresca/Maximum Shear Stress Criterion

In a plastic working process where the stresses may not be uniaxial, it is necessary to apply a yielding criterion for the stress system which is acting. The Tresca yield criterion hypothesis is related to the maximum and minimum principle stresses, and defines yielding as occurring at a critical value of the maximum shear stress, τ_{max}.

$$\tau_{max} = \frac{\sigma_1 - \sigma_3}{2} = \pm k \tag{6.1}$$

σ_1 = maximum principle stress
σ_3 = minimum principle stress
k = constant

For a tensile or compression test, $\sigma_2 = \sigma_3 = 0$.
Then $\sigma_1 = 2k = \sigma_y$
In general, $\tau_{max} = \dfrac{\sigma_y}{2}$

6.2.2 Von Mises Criterion (Distortion Energy Theorem)

The Tresca criterion takes into account only σ_1 and σ_3, while the Von Mises criterion considers all three principal stresses. Plastic flow in the Von Mises formulation occurs with the following criterion:

$$\left\{ \frac{1}{2}[(\sigma_1 - \sigma_2)^2 + (\sigma_2 - \sigma_3)^2 + (\sigma_3 - \sigma_1)^2] \right\}^{\frac{1}{2}} = \sigma_y \tag{6.2}$$

The two criteria when compared show a difference of 15% between them.

6.3 Work Hardening of Metals

Metals work harden during plastic deformation and the yield stress increases with strain. A power law relationship between true stress and strain is obtained with the following form:

$$\sigma = S_1 \varepsilon^n \tag{6.3}$$

n = strain hardening exponent
S_1 = strength coefficient

n can vary between 0.1 and 0.4. Materials with $n = 0.4$ are those with high strain hardening ability e.g. stainless steel.

It is convenient in calculations to consider a material in which $n = 0$ i.e. without strain hardening. Then σ_y = constant and the stress strain curve after yielding becomes a line parallel with the ordinate.

6.4 Microplasticity and Grain Size

The plastic working of metals is related to processes involving the movement of dislocations. These move in metals under stress. The structure of metals is crystalline and a crystal grain is defined as a region contained by a boundary within which the crystallographic planes are uniformly oriented. A grain boundary is located at the interface between crystalline areas of different orientations. Grain size may be measured from the pattern of boundaries observed by microscopy. The yield stress of a metal varies with grain size and increases with a decrease of crystal dimension, according to the Hall Petch equation:

$$\sigma_y = A + Bd^{-\frac{1}{2}} \tag{6.4}$$

d = grain size
A, B are constants

Deformation processes in polymers occur by different mechanisms not connected with dislocations. Ceramic materials are crystalline and dislocations are present in the structure, but ceramic materials have little or no plasticity and this is related to the immobility of the dislocations. The processing of ceramics is therefore limited principally to powder operations followed by sintering.

6.5 Rate and Temperature Effects in Metal Forming

The yield stress is also dependent on temperature and strain rate. For the latter, the following power law can be applied

$$\sigma_y = B\dot{\varepsilon}^m \tag{6.5}$$

B is a constant
$\dot{\varepsilon}$ is the strain rate
m is the strain rate sensitivity

The yield stress is inversely dependent on temperature, and this is more significant than the strain rate dependence. In steel, a change in strain rate by a factor of 10^3 would be required to compensate for a temperature rise of approximately 50 °C.

6.6 Plastic Working of Metals

Metals are worked plastically by different processes to give final shapes required in technology. The methods include forging, rolling, drawing, extrusion, wire drawing. Some analyses will be given.

6.6.1 Wire Drawing

A wire drawing operation is shown in Figure 6.1. This shows a rod of cross sectional area A_o being reduced to A_1 in a die. The drawing force on the wire at the die exit is F_d and the drawing stress is σ_d. The angle of the die is α. The wire of original cross section A_o is reduced by the stresses acting on it due to the reaction of the die wall to the drawing stress σ_d. The maximum reduction which can be obtained is limited by the stress on the wire at exit. When the exit wire begins to deform plastically, it may fail by necking and a limit has been reached to the reduction. This fixes the maximum value of σ_d when

$$\sigma_d = \sigma_y \tag{6.6}$$

From this, it is possible to obtain the maximum reduction possible in wire drawing. If the work balance is calculated up to σ_d, the drawing stress, the wire on the inlet side moves l_0 with a cross section of A_o and on the outlet side, it moves l_1 with a crosssection of A_1. The work on the outlet side is

$$F_d l = \sigma_d A_1 l_1 = \sigma_d V \tag{6.7}$$

The work per unit volume is σ_d.
σ_d is the work under the stress strain curve up to the strain at exit.

$$\sigma_d = \int_0^{\ln A_0/A_1} \sigma \, d\varepsilon \tag{6.8}$$

This has to be multiplied by $1/\eta$ where η is the efficiency of the operation. Then

$$\frac{1}{\eta} \int \sigma \, de = \sigma_y \tag{6.9}$$

$$\sigma_d = \frac{S_1}{\eta} \int^\varepsilon \varepsilon^n \, d\varepsilon = \frac{S_1}{\eta(n+1)} \varepsilon^{(n+1)} \tag{6.10}$$

$$\sigma_d = \frac{S_1 \varepsilon^n}{\eta(n+1)} \cdot \varepsilon = \frac{\sigma_y \varepsilon}{\eta(n+1)} \tag{6.11}$$

$$\varepsilon = \eta(n+1) \tag{6.12}$$

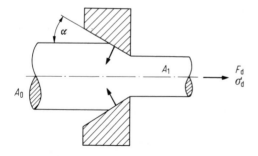

Fig. 6.1 Wire drawing.

Thus, the maximum strain can be evaluated which gives A_o/A_1. It is related to the efficiency η. Metals which have a high strain hardening exponent can be drawn to bigger reductions in one pass.

6.6.2 Rolling

An analysis of rolling examines the distribution of stress between the rolls in a rolling operation, Figure 6.2a. The analysis proceeds to calculate what is termed the "friction hill", the pressure distribution on the rolls, and the roll torque.

Figure 6.2(b) shows the area of contact between rolls in a rolling mill and a steel slab of initial thickness h_1. The slab is rolled to a thickness h_2. The projected length of the arc of contact = L.

$$L = BC = BD \cdot AB.$$
$$BD = 2R \text{ where } R = \text{ roll radius.}$$
$$AB = (h_1 - h_2)/2$$
$$L = \sqrt{R(h_1 - h_2)} = \sqrt{R\Delta h}$$

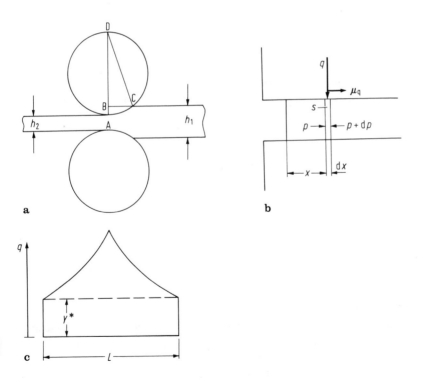

Fig. 6.2 (a) Rolling geometry. (b), (c) Friction and the roll pressure distribution.

The projected area of contact is L multiplied by the width of the rolled material. To calculate the roll force, friction at the rolls must be computed, since this introduces an additional compressive force. An approximate calculation can be made using as a model of the process, a slab compressed between plane plates. If μ is the coefficient of friction between slab and plate,

$$2\mu q\, dx = h \cdot dp$$
$$q = \text{vertical compressive stress.}$$
$$p = \text{mean pressure acting on left hand side.}$$

The condition for plastic yielding of the segment is

$$q - p = Y^* \quad \text{where } Y^* = 1.15\sigma_y$$

Then $\quad dq = dp \quad$ since Y^* is a constant

and $\quad 2\mu q\, dx = h\, dq$

$$\frac{dq}{q} = \frac{2\mu}{h}\, dx$$

Hence

$$q = q_0 e^{\frac{2\mu}{h}x}$$

q_o = value of q for $x = 0$, i.e. at end of slab.
At $x = 0$, the horizontal pressure p vanishes
i.e. $q = Y^*$

$$q = Y^* e^{\frac{2\mu}{h}x}$$

Since the vertical pressure in the model must be considered as symmetrical about the center, and replacing the curves for the distribution of stress by straight lines,

$$q = Y^* \left(1 + \frac{2\mu}{h}X\right)$$

If the length of the slab = L, the maximum pressure at the centre is given by

$$q_{max} = Y^* \left(1 + \frac{\mu L}{h}\right)$$

Therefore, the mean pressure $\bar{q}$ is given by

$$\bar{q} = Y^* \left(1 + \frac{\mu L}{2h}\right)$$

$\dfrac{\mu L}{h}$ is an important parameter.

The calculation of roll torque can be performed approximately, again assuming that the roll pressure is distributed symmetrically around the centre of the arc of contact, Figure 6.2(c). If L is the projected length of the arc of contact, the lever arm of the vertical roll pressure with respect to the centre of the roll is $L/2$. The total torque upon both working rolls is FL.

In effect, the lever arm of the roll pressure is less than $L/2$, but it is suggested that using the value $0.5L$ is safer when the value of the torque is used for choosing the dimensions of the driving motor or the drive.

6.7 Polymer Processes

Polymers cannot be poured into moulds because of their high melt viscosity, and the processes to be described are dependent on viscous flow. The coefficient of viscosity relates the shear stress to the rate of shear strain in the Newtonian relationship

$$\tau = \eta \dot{\varepsilon}$$

τ is the shear stress
$\dot{\varepsilon}$ is the rate of shear strain
η is the coefficient of viscosity

6.7.1 Classification of Polymers into Processes

The following is a classification of processes for different types of polymer.

Thermoplastic materials are processed by Calendering, extrusion, injection moulding, blow moulding.

Thermosetting materials are processed by simple moulding techniques.

Rubbers are processed by extrusion and moulding.

6.7.1.1 Screw extrusion

This process shapes a molten polymer by forcing it through a die, by means of one or more screws rotating inside a heated barrel. The form of the die determines the initial shape of the extrusion. Figure 6.3 shows a single screw plasticity extruder.

Solid feed material is in the form of a granules or powder. This material is compressed and transported into the machine. Melting occurs by external heat. The metering section is of constant depth, intended to control the output of the machine and generate the heat necessary, delivering pressure and mixture of the melt.

A perforated breaker plate is situated at the end of the screw. The melt is forced through this. It holds back unmelted polymer, metal particles, or other foreign matter and evens out temperature variations.

The pressure is in the range of 40 MN m^{-2}.

6.7.1.2 Flat film and sheet extrusion

In this method, the melt flow is spread laterally to produce uniform sections. The melt is distributed by a manifold and passed through a narrow taper channel which controls the thickness (Figure 6.4). The film is subsequently wrapped round a water cooled roller.

6.7.1.3 Pipe extrusion

The equipment for pipe extrusion is shown in Figure 6.5. A screw forces the molten plastic between an annular outer die and a central cylindrical piece called a torpedo. This is held by 4 spiders and may be a source of introducing imperfections in the tube. The calibrator controls the outer diameter, and the tube is cooled by a water bath.

6.7.1.4 Tubular film extrusion

This process produces tube by film blowing, Figure 6.6. The production is vertically upward, the extruder turning the melt through a right angle, when it is expanded by internal air pressure to form a bubble. This is structured in the direction of flow

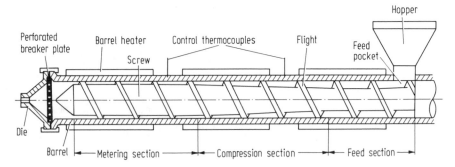

Fig. 6.3 Single screw plastics extruder (from Fenner 1979).

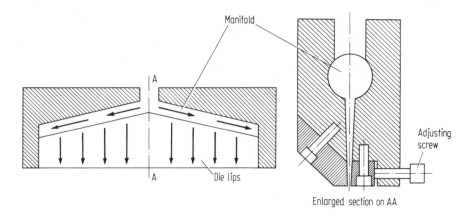

Fig. 6.4 Flat film and sheet extruder (from Fenner 1979).

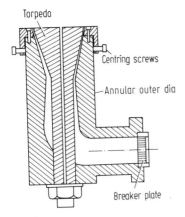

Fig. 6.5 Pipe extruder (from Fenner 1979).

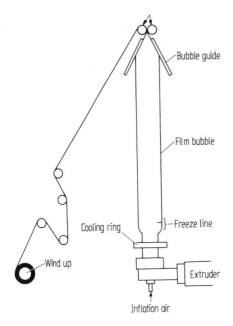

Fig. 6.6 Tubular film-blowing process (from Fenner 1979).

by powered nip rollers. These close the bubble and the flattened film is wound on a reel.

6.7.1.5 Injection moulding

An injection moulding machine is shown in Figure 6.7. This machine has an injector of the single screw type. Molten polymer is accommodated between the end

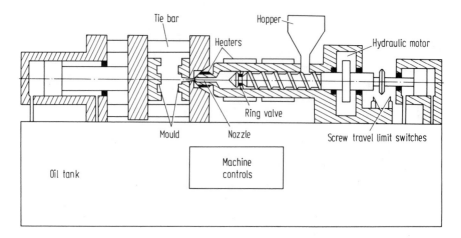

Fig. 6.7 Injection moulding (from Fenner 1979).

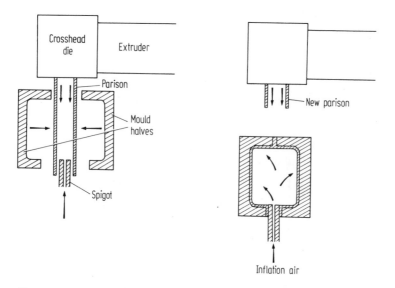

Fig. 6.8 Blow moulding (from Fenner 1979).

of the screws and the injection nozzle by halting the screw and moving it back. The polymer is then injected into the mould by moving the screws forward by the hydraulically applied axial load. Backward flow along the screw channel is avoided by a ring valve on the tip of the screws.

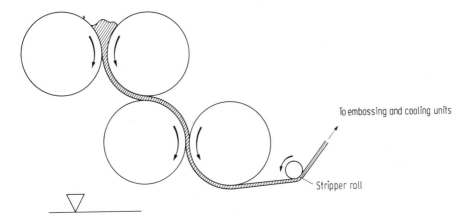

Fig. 6.9 Calendering (from Fenner 1979).

6.7.1.6 Blow moulding

This process makes hollow articles, e.g. bottles and containers. The movement is vertically downwards from a cross head of a narrow thick-walled tube. The mould surrounds the tube. The melt is called a "parison". A spigot, incorporating an inlet for compressed air, then blows the parison into contact with the water-cooled mould. The moulded part is then trimmed of excess material (Figure 6.8).

6.7.1.7 Calendering

This process uses horizontal rolls to produce a continuous sheet. Figure 6.9 shows a 4-roll unit in which the speed and temperature of the rolls are independently controlled. Molten material is supplied by a screw extruder and is fed to the calender as continuous strip. Sheet is removed finally by a high speed stripper roll.

References

Astarita G, Nicolas L (Eds) 1983. Polymer Processing and Properties. Plenum.

Avitzur B (1968). Metal Forming: Processes and Analysis. McGraw-Hill, New York.

Boer CR, Rebelo N, Rydstad H, Schroder G (1986). Process Modelling of Metal Forming and Thermomechanical Treatment, Springer.

Dieter GE (1986). Mechanical Metallurgy. McGraw Hill.

Fenner RT (1979). Principles of Polymer Processing, Macmillan.

Hill R (1956). Mathematical Theory of Plasticity, Oxford.

Hunt VD (1983). Industrial Robotics Handbook. Industrial Press Inc. N.Y.

Kusiak A (Ed.). Flexible Manufacturing Systems. Methods and Studies. N. Holland.

Keeney RL (1982). Decision Analysis: an Overview. Operations Research 30, 5:803.

Keeney RL, Raiffa H (1976). Decisions with Multiple Objectives, Preferences and Value Trade offs. John Wiley N.Y.

Kochhar AK, Burns ND (1983). Microprocessors and their Manufacturing Applications. Edward Arnold.

Kowalik JS (1986). Knowledge Based Problem Solving. Prentice Hall.

Krockel H, Reynard K, Stven G (Eds) (1986). Factual Material Data Banks. CEC Workshop. Petten.

Miller WE, Automation of Metallurgical Processes. An Overview. Drive Systems Dept. GEC Salem.

Moffat DW (1987). Handbook of Manufacturing and Production, Management formulas, Charts and Tables, Prentice Hall.

Muller RS, Kamins THI (1986). Device Electronics for Integrated Circuits 2nd Ed. John Wiley.

O'Brien JJ (1970). Management Information Systems, Van Nostrand Reinhold.

O'Shea T, Eisenstadt M (1984). Artificial Intelligence. Harper and Row.

Orowan E (1955). Lecture Notes. MIT Cambridge, Mass.

Pearson JRA (1983). Computational Analysis of Polymer Processing. Appl. Sci. Publ.

Pittman JFT, Zinkiewicz OC, Wood RD, Alexander JM (Eds) (1984). Numerical Analysis of Forming Processes. John Wiley.

Plastics and Rubber Industries (Eds) (1987). Plastics and Polymer Processing Automation. Noyes Data Corp. USA.

Tadmor, Z, Gogos CG (1979). Principles of Polymer Processing, John Wiley.

Ott ER (1975). Process Quality Control McGraw-Hill.

Palm WJ (1986). Control Systems Engineering. John Wiley.

Parent M, Laurgeau C (1985). Robot Technology Vol. 5, Logic and Programming (English Transl.) Prentice Hall.

Pidd M (1984). Computer Simulation in Management Science, John Wiley.

Rydz JS (1986). Managing Innovation. Ballinger Publ Co.

Shingo S. (1985). A Revolution in Manufacturing. The SMED System. Productivity Press, Standford.

Sol HJ, Takkenberg CAT, DeVries P (1985). Expert Systems and Artificial Intelligence in Decision Support Systems. Proc 2nd MiniEuroConference, Luuton Netherlands.

Tijnaelis D, McKee KE (1987). Manufacturing High Technology Handbook, Marcel Dekker.

Walker TC, Miller RK (Eds) (1986). Expert Systems. An assessment of Technology and Applications. SEAI Technical Publications, Madison GA 30650.

Zinkiewicz OC, Morgan K (1983). Finite Elements and Approximation, 1st Ed. John Wiley.

Zuboff S (1988). The Future of Work and Power. Basic Books, N.Y.

Chapter 7

Process Selection and Control/Computer Applications/ Electronics Manufacturing Processes/Modelling of Processes

7.1 General

In this chapter, an outline is given of some of the procedures required to select a process and to establish the methods required for its control. Some description is given of different techniques in planning production including flexible manufacturing systems, group technology and integrated manufacturing production. A short survey is given of aspects of information technology, computers and microprocessors, their components, and methods available for their manufacture.

7.2 Costing

Cost of a manufacturing process must be calculated in introducing a new system and must be related to the materials and to steps in the process. Cost effectiveness has been suggested to mean that a product must be offered for sale at a marketable price and with the assessment that it will perform its function efficiently throughout its useful life. This concept leads to the term "Life Cycle Cost (LCC)".

The general analysis of production costs involves procedures for determining the prices of equipment, tooling and materials handling. A market analysis is also made. A consumer evaluation should include factors like quality, reliability, dimensional stability, accuracy.

There are comparatively simple methods for costing, one approach being based on annual production volume. The unit cost is determined by the investment, the price of materials and the productivity. Figure 7.1 shows an example. This is for a cost calculation based on systems involving manual, automatic and robotic manufacturing methods. The graph relates cost with production volume based on the particular manufacturing method selected.

7.3 Technical Cost Modelling

Technical cost modelling introduces the properties of the product in addition to the costs determined by materials and production, as in 7.2. The properties, together with process cost will interrelate to determine consumption.

Typical cost modelling involves a process flow diagram, which includes the

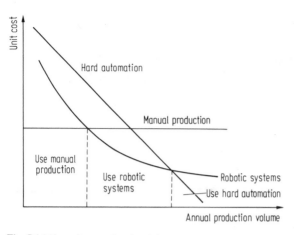

Fig. 7.1 Life cycle cost. Reprinted from PM Francis and TH Heizerman in Manufacturing High Technology Handbook, Eds D Tijnaelis, K. McKee, 1987 p. 344 by courtesy of Marcel Dekker Inc.

kind of capital equipment involved in the process followed by the direct costs. These may include the following:

1. The primary material, e.g. Aluminium, Steel
2. The process materials e.g. die materials, tools etc.
3. Labour
4. Energy
5. Cost of capital
6. Overheads

The cost of a process, also termed process cost modelling, utilizes computer models to simulate production costs and analyses the effect on cost of factors such as production volume and the yield (e.g. in a casting process).

7.4 Decision Analysis Techniques and Materials Selection

A growing number of alternative materials is presented for any process, and complicates decision making. One decision making process is multi-attribute utility analysis (MAUA). A metric of various combinations of characteristics is prepared. The metric is called a utility. The relationship between the value of a set of characteristics and the utility of that set is called the utility function.

Figures 7.2a and b give graphs obtained from multi-attribute utility analysis. The graphs are called Iso-Utility lines. Both graphs are related to the cost of a process or a part. In Figure 7.2a, the cost of an automobile bumper is related to its weight. Figure 7.2b shows cost of a tool related to cutting speed. For calculations of alternative processes, the new system proposed will be attractive if it falls below the iso-utility line. As an example, in Figure 7.2a, system 4 would not be attractive.

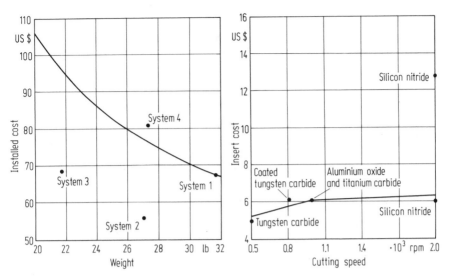

Fig. 7.2 (a), (b) Multi-attribute Utility Analysis, from Advanced Materials and the Economy by Joel P. Clark and Merton C. Flemings, 1986, Scientific American.

Figure 7.2b shows figures for Si_3N_4 as a tool material. Initially the cost is high, although it operates at a high cutting speed. The analysis shows what the cost of a part manufactured from Si_3N_4 needs to be to make it competitive in price as a cutting tool and this must be about $ 6 .

In the process of making a decision, a utility function of the decision maker is estimated by an analyst. This is used to gain information of the criteria employed by the decision maker and it can make predictions of the decision makers actions under certain conditions. One technique of analysing the data is called the Keeney-Raiffa interview technique. The analyst uses a questionnaire for the decision maker which is designed to represent situations similar to the problem being solved. The decision makers preferences are obtained in an analytic form and the utility function is examined. The treatment of alternatives in the metric can be made.

7.5 Materials Availability and Selection/Data Banks

The selection of a method must be related to the properties of materials, their availability and cost. Data banks and source books exist for materials and the selection of materials has been the subject of computer investigations. A short listing of data banks is given in Table 7.1. A listing of source books is given at the end of this chapter.

A data-base, a set of data stored on a computer, can be accessed by the users in a selective manner, and in a reasonable time frame. A soft-ware system allows communication between the user and the data-base.

Table 7.1 Data banks for materials.

Name	Material	Source
METADEX	METALS	Metals Society (ASM International) (World's largest) Computer file on Metallurgy)
EMA (on line equivalent of Engineered Materials Abstracts. Available on Dialog, ESA-IRS and Orbit Search Service	Polymers, Ceramics Composites	Metals Society (ASM International)

7.6 Information Technology and Production Processes

Information technology in production processes is an important field which touches on subjects like inventory analysis, materials flow, the diagnosis of manufacturing problems and their correction. Some of the processes which are involved based on the computer are:

Data preparation
Data retrieval
Systems analysis
Quality control

Production processes make use of computer science from fields such as Artificial Intelligence and Expert systems. Management innovation increasingly relies on these sources.

7.7 CAD-CAM/The Computer as a Machine

Computer aided design and computer aided manufacturing are a part of present technology. Schon remarked that industrial innovations do not normally cause sweeping changes but computer aided manufacturing is a revolutionary innovation. It is therefore necessary to pay attention to the computer as a machine, as well to study social acceptance of the computer and its influence on manufacturing. Reference should be made to the studies of Shoshana Zuboff and Charles Handy.

The hardware normally available for CAD is a computer, graphics display terminals and keyboards. Different software is available.

CAM is a rapidly growing field which touches on different areas to be discussed later in this Chapter such as Flexible Manufacturing and Integrated Manufacturing Systems (IMS). A general description of CAM starts with a direct or indirect computer interface, allowing planning, management, and control of the operation of a manufacturing plant. The direct approach requires sensing devices, e.g. thermocouples or strain gauges, which are directly linked through an analogue device to give a digital input to the computer. Monitoring or control of the process is made through a control algorithm.

In the indirect approach, the computer is used for production scheduling, process planning, materials selection, standards etc. Non-destructive testing is an important field for computer application. It is used in production processes with the techniques of acoustic emission, eddy currents, ultrasonics etc. and relies on high speed acquisition of data and its processing.

All the techniques which go into the achievement of quality in production must be studied.

7.8 The Computer in Manufacturing Technology/Turing's Machine

In his analysis of Western Culture and the Computer Age, Bolter discusses the Turing Machine. In the introduction to this book, A.J. Ayer draws attention to the description by Turing in 1936 of symbolic logic as a mechanical procedure. In Engineering, the accepted sense of a machine was a device for producing power to do work. Turing's machine did not perform work in this sense. In terms of the new technology, it was a processor. A computer processes information. Turing's machine moved its marker back and forth along its tape, examining, erasing and writing symbols as it applied its rules of operation. It is a logic machine and consists of two parts. One part is a finite set of operating rules. The second part is a tape of unlimited length upon which changeable information can be stored.

The tape is divided into cells, each of which may contain one symbol and there is a marker to indicate which cell is being inspected at any one moment. The machine moves the marker back and forth along the tape, examining, erasing, and writing symbols in applying its rules of operation. Turing's machine was in fact a computer. It replaced symbols one at a time according to a finite set of rules.

7.9 The Von Neumann Machine

Von Neumann's machine is the proto-type modern computer. It is made up of a central processing unit (CPU) where Arithmetic and Logic operations are performed and a memory (storage unit) where instructions and data are kept waiting their turn in the processor. Coded information is read into the CPU where it is computed and the CPU then writes the information back.

7.10 Microprocessor

A microprocessor is a synonym for a microcomputer, i.e. CPU and peripheral unit interfaces. Basically it is a chip transistor, which is a large scale (LS1) or very large scale (VLS1) integrated circuit. Processes and devices in electronics are described in the following.

7.11 Processes in Electronics

Processes in electronics involve the production of materials in high states of purity e.g. Silicon, the deposition of thin films on prepared surfaces, selected etching or cutting to form separate units of film, and joining processes to make contacts between these units. Some of the techniques have already been described e.g. CVD, ion implantation, solid state bonding. The processes are in a state of rapid change as further development takes place to produce more units on a chip and as other materials become of interest in circuitry e.g. polymers.

Silicon crystals used in circuitry are doped with impurities of p or n type. These alter the densities of the two types of charge carriers. The n type carriers donate electrons. The p type carriers contribute to hole conduction and this is termed an acceptor input. In circuitry, the important properties arise from the interaction between adjacent semi-conductor materials having different densities of the two types of charge carrier. Some description will be given of the processes which are employed in electronics for silicon crystals.

7.12 Growing Silicon Crystals

Figure 7.3 shows the principle of growing a single crystal of silicon, using the Czochralski technique. A crucible of molten silicon is held at a controlled temperature and a seed crystal of silicon having a prescribed orientation is lowered into the melt. It is then withdrawn at a predetermined rate, while rotating. It draws with it a crystal of silicon which has grown from the melt. This growth occurs at an interface which is formed slightly below the melt surface. The seed crystal which was initially cut with a required orientation determines the growth of the bulk crystal in a set direction.

The cylindrical crystal which is produced may be in the size range of 10 cm in diameter. It is sliced into wafers for device production.

7.13 Device Technology

Devices are formed on the surfaces of the silicon after cutting and polishing. The dopant atoms, required to form p-or n-type material, are introduced into the silicon in exposed regions of the surface, while an oxide film protects the remaining surface. Joints between units are made of metallic conductors. The protecting film is made by thermal oxidation, while the selective removal of the SiO_2 is by photolithography.

The ability to form smaller device units, and thus to pack more devices into a unit area is also dependent on the technology of forming the gap between units. The prevalent techniques use photo-resist and photo-lithography but these are subject to competition by laser techniques, employing as an example excimer lasers.

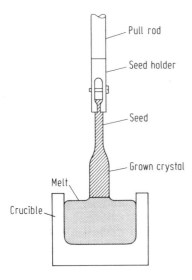

Fig. 7.3 Czochralski single crystal technique.

Some of the more common steps in what is termed ''planar technology'', are described in the following.

7.13.1 Planar Technology

Planar technology involves the preparation of surfaces on wafers cut from single crystals, using different techniques to form devices. A device is a unit performing a function in an electronics system, and is based on electron flow dependent on semi-conductor properties and behaviour of the device in an imposed electric field. Examples of devices are transistors and capacitors.

7.13.2 Thermal Oxidation and Photolithography

The SiO_2 layers are formed by different processes, such as CVD and thermal oxidation, the latter being performed in an oxygen stream at 850–1100 °C. The selective removal of the oxide film is made by using a light sensitive polymer, called a ''resist''. This is spread on the oxide surface, e.g. by spinning, and dried. A mask, which is a photographic negative with transparent areas is placed over the photoresist and the whole is exposed to ultra-violet light. The molecules of a negative resist become polymerised (cross-linked), and in a positive resist, the molecular bonds are broken. The unaffected areas are selectively dissolved using a solvent such as trichlorethylene, and the oxide film in these areas can be removed by HF. This is followed by removal of the resist.

7.13.3 Integrated Circuit

An integrated circuit (IC) is prepared by forming several semi-conductor devices on a silicon chip and connecting them together. The interconnection is made normally by depositing Al or polycrystalline Si.

7.13.4 Transistor

The transistor, formed by a junction between p and n type semi-conductors was the initiating idea in modern electronics. Under bias, the p region is positive with respect to the n region. The junction conducts current by holes injected from the p region and electrons from the n region. As the voltage increases, the current rises rapidly. Under reverse bias, the currents are smaller, but can be increased by different means of supply of minority carriers. Modulation of current flow properties can be affected.

7.13.5 Packaging in Electronics

Packaging in electronics refers to the material and the means for mounting a finished chip. The connection between each chip or high density interconnect (HDI), is a field of development related to the scale for fine conductor lines in a multilayer structure. At close line spacing, conductors must be resistant to electrochemical migration, or very well protected against small amounts of atmospheric moisture.

7.13.6 Electron Beam Lithography

Patterns on wafers can be obtained by electron-beam sources and line widths can be in the region of 0.1 μm. The electron beam is swept over the surface at right angles to the motion of the wafer.

7.13.7 Epitaxy

A silicon film which is crystallographically a single crystal can be deposited on the surface of a single silicon crystal, e.g. by CVD. The epitaxial growth process is related to one nucleation event and growth maintaining an orientation of the new crystalline film in a fixed relationship with the original crystal.

The deposition of Si from the vapour on amorphous SiO_2 is normally a process in which multiple nucleation occurs. The polycrystalline film is used as a conductor or electrode in integrated circuits.

7.13.8 Monolithic Integrated Circuit (MIC)

This is a solid block of electronics equipment with no connecting wires. Oxidation of a crystal is performed at high temperature to form a stable oxide on the surface.

This acts as a mask and openings through this are made in selected locations, which allow impurities to be introduced into the silicon by diffusion or ion implantation.

7.14 Manufacturing Organisation

Different possibilities for organising manufacturing processes have resulted from the development of computer science. Originally manufacturing processes were operated sequentially, i.e. the parts to be manufactured were shaped progressively in specialised departments of factories and finally achieved their required form.

One of the newer concepts is to be found in cellular manufacture. The cell is a production unit which contains the equipment used in several operations. The manufacturing process is organised by computer with a programme which integrates the different operations. The manufacturing unit has now become a cell which houses the different procedures, rather than the traditional production lines which were based on sequential single operations. Such methodology has been found to optimise labour, production control, quality control, machinery maintenance etc. A plan of a manufacturing cell is shown in Figure 7.4.

Modern manufacturing systems are also concerned with "flexibility". Flexible systems may be described as those which react quickly to changes in product demand and product design. This is described further in Section 7.15.2.

CAD (computer aided design) provides a tool for engineers to develop new products on a computer.

7.14.1 Analysis of Types of Production Job Shops

Production job shops in the process industry would include machine shops, foundries, plastics moulding plant, forge etc. Black differentiates between shops with a limited through-put, and mass production shops with a large flow of parts. The latter systems are very specialised. Changes in design of the process or in the process itself, are expensive and hence are generally avoided.

For machining operations, to allow flexibility, PLC's (programmatic logic controllers) are used. In addition the machinery is based on modular units, that can accomplish a function, rather than produce a specific part.

7.14.2 Analysis Methods for Selecting Processes

Selection of Materials was discussed in Section 7.5. In materials processing, a growing number of materials are becoming available and associated with these are a growing number of processes. While materials data banks exist, process data banks are not available, because of the complexity of the processes and the difficulty of coding information for a computer. A large number of different procedures have been described for different processes. A reasonably good example of a production procedure is the following for assembly manufacture related to vehicles.

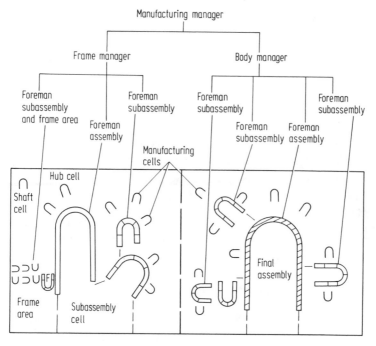

Fig. 7.4 Manufacturing cell. Cellular Manufacturing System (CMS) is composed of manufacturing cells, subassembly cells and final assembly lines. Cell is often in U shape to enable the worker to move from machine to machine. In the manufacturing cell, processes are grouped according to sequence of processes and the operations needed to make a group of family of parts or products, (from Black 1988, with permission of ASME, United Engineering Center, New York).

1. Prepare engineering drawings for assemblies.
2. Outline the manufacturing processes.
3. Design tooling.
4. Order tools.
5. Prepare materials specifications.
6. Prepare test assemblies and test.
7. Test the weight of assemblies.
8. Accept or improve procedures.

A more analytical approach would be the following:

1. Organise data preparation and retrieval system for the materials and processes involved.
2. Use systems analysis and define the operation.
3. Write programmes for the entire process involving production algorithm, management information and quality control.
4. Network the computers.

7.15 Integrated Manufacturing Systems (IMPS)

The term Integrated Manufacturing Production System (IMPS) has developed from the CMS (Cell Manufacturing System) (see Section 7.14) and the cell developed from the production jobbing shop. Manufacture now groups together units of differing functions under control of a computer to produce a finished or semi-finished item. The cell uses robots and automation, together with computing to solve problems. If the system can re-organise itself rapidly to changes of demand and product design it has acquired the status of being a flexible one.

From this understanding of organisation, the term "integrated system" is applied, based on the overall methodology of the cellular manufacturing system.

7.15.1 Group Technology

Group Technology collects components to be manufactured and organises families of these. These families are based on components having a similar design or manufacturing sequence. The family of components is then processed by a group of machines in the cell and these represent a set of processes e.g. cutting, forging, machining, surface treating. The machines in the cell have an order which defines the manufacturing sequence, replacing the operations of individual job shops.

7.15.2 Flexible Manufacturing Systems

Flexibility in manufacturing systems means adaption to change, both in production volume of parts, and design of components. It implies a rapid response to changing requirements in the system. In the choice between powder metallurgy part or a forging, as an example, some of the decisions to be made would be as follows:

1. Which method of financial judgement would be used?
2. Is the chosen process capable of using robotics?
3. Use of computer and information systems for the two processes.
4. How is the control of quality to be instituted?
5. How reliable is the product manufactured by the two processes?
6. Competitive prices.
7. What new aspects of the technologies involved must be considered?
8. Newer materials.
9. Can the system be more cheaply operated by an entirely different technology, e.g. casting?

7.15.3 Changes Involving Development of New Systems and Materials

The systems should be examined for possible changes, which might require serious production re-organisation, e.g. (1) changes from a sheet metal forming system to a glass fibre reinforced polymer system; (2) Change from a metal to a ceramic component or metal matrix composite.

7.16 Automation of Materials Processing

The processing of materials, and the introduction of new techniques in automation, has brought with it further new requirements in process control. The current development of process modelling, and broader computer applications including artificial intelligence are important. Process models and artificial intelligence are described separately in 7.17.

7.16.1 Robots and Robotics

Industrial robots have three principal components. A robot usually has one or more arms situated on a fixed base, and capable of moving in several directions. There is a manipulator which holds a tool, or the part which needs to be worked. The robot has a controller, which gives the detailed instructions. As an example, in robot welders, a sensing device is a necessary part of the machine where a requirement for keeping the welder on track is important. Robotic systems are readily programmed and interfaced with equipment. An example is that of a robot spot welder. In the automobile industry welding must be performed on components which deviate from a fixed location of the welding machine. The welding robot must be adapted to compensate for the deviation. A method for controlling robot spot welding of car bodies involves a high speed arithmetic LSI. This makes it possible to operate a 6-axis mechanism easily with a linear or circular control and in combination with sensors and sophisticated software.

7.17 General Aspects of Process Modelling

A model of a system is its mathematical or symbolic representation. Materials processing systems are presently capable of being quantitatively described and in the first chapters of this book, some of the mathematics were given. Equations were written relating to mechanical working, plastic deformation, solidification processes and joining. Starting from the quantitative description, there are different routes which may be taken to model the system. One possibility is to solve the equations analytically by a mathematical method. The solution is then used as a means of predicting the change of a parameter in the system for changes in the variables. An example would be for the temperature distribution in a plate which has a welding arc moving over its surface.

Starting from a quantitative description, a programme can be made which models the system for a computer. This model can accomplish important functions. It allows the possibility of compressing real time so that years of operation can be studied in a few minutes.

The process can also be slowed down and studied in detail. The computer can supply solutions to many of the problems in manufacturing technology and it can provide the answers more economically than with tests in the laboratory or by using production facilities. Tests and experimental research are still required, but the computer can be a powerful tool in directing and analysing these.

Where analytical solutions prove difficult, as for example in heat transfer problems with complex geometries, numerical methods can be used. These involve the use of techniques such as Finite Differences and Finite Elements.

Computer techniques are emerging as an important field for process understanding and development in engineering. Applications are being made successfully in welding, casting, metal working processes, powder processes and others.

The finite difference and the finite element methods will be briefly described in modelling solidification processes.

7.17.1 Solidification Modelling

Consider first solidification problems. Modelling techniques allow processes, such as the casting of metals, to be analysed so that solutions of a number of problems can be obtained without resorting to lengthy empirical methods. Design problems can be solved such as the location and size of risers, and the dimensions of gating systems.

There are two directions for modelling solidification, termed macro- and micro-techniques. In macro-modelling, the heat transfer equations are solved to determine the location of isotherms in relation to time and thus to map the progress of solidification. In micro-modelling, the evolution of micro-structure is simulated using expressions for nucleation kinetics, growth of dendrites, and growth of eutectics. The applications are also related to the melting of surfaces e.g. by lasers, the modelling of film growth, rapid solidification etc.

7.17.2 Finite Difference Method

For three dimensions, the heat transfer equation can be written

$$\frac{\partial T}{\partial t} = \alpha \left(\frac{\partial^2 T}{\partial x^2} + \frac{\partial^2 T}{\partial y^2} + \frac{\partial^2 T}{\partial z^2} \right) + L$$

α is the thermal diffusivity $= K/\rho C$
K is the thermal conductivity
ρ is the density
C is the specific heat
L is the latent heat of solidification.

In order to find a solution to the above equation and avoid the problem of accounting for latent heat, it can be rewritten in terms of the enthalpy H

$$\frac{\partial H}{\partial t} = \frac{K}{\rho} \left(\frac{\partial^2 T}{\partial x^2} + \frac{\partial^2 T}{\partial y^2} + \frac{\partial^2 T}{\partial z^2} \right)$$

$$\partial H = C \partial T$$

$$\frac{\partial H}{\partial t} = C \frac{\partial T}{\partial t} + \frac{\partial L}{\partial t} = C' \frac{\partial T}{\partial t}$$

This incorporates the latent heat as an increase in the apparent specific heat C. The heat flow equation is next written in finite difference form. In the finite difference method, a function, e.g. T_m is known at regular intervals where $m = -2, -1, 0, 1, 2$. Forward differences are written

$$\frac{\partial^2 T}{\partial x^2} = \frac{T_{i+1}^n - 2T_i^n + T_{i-1}^n}{(\Delta x)^2}$$

$$\frac{\partial H}{\partial t} = \frac{H_i^{n+1} - H_i^n}{\Delta t}$$

$$\frac{\partial T}{\partial t} = \frac{T_i^{n+1} - T_i^n}{\Delta t}$$

The geometry of the casting is divided into small but finite sized elements, Figure 7.5.

Δx is the space division in Figure 7.5.
Δt is the time increment

The superscript denotes a time interval. There are two methods adopted for solutions. In the explicit method, the element itself and the adjoining elements are considered isothermal at the initial known temperatures. The finite difference equation has the following form

$$\frac{H_i^{n+1} - H_i^n}{\Delta t} = \frac{K}{\rho} \cdot \frac{T_{i+1}^n - 2T_i^n + T_{i-1}^n}{\Delta x^2}$$

The only unknown value is H_i^{n+1}. It is calculated and the relationship between H and T is then used to calculate T at each nodal point. This simple formula is repeated to determine the T's at each time increment to follow the progress of solidification.

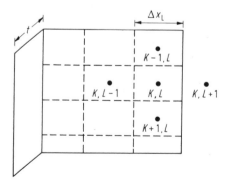

Fig. 7.5 Elements of finite difference mesh.

In the implicit method, an isothermal relationship is taken at the final unknown temperature. The equations are written

$$\frac{H_i^{n+1} - H_i^n}{\Delta t} = \frac{K}{\rho} \cdot \frac{T_{i+1}^{n+1} - 2T_i^{n+1} + T_{i-1}^{n+1}}{\Delta x^2}$$

The equations for all the nodal points and for all the time steps need to be written and solved simultaneously. The method is more accurate than the explicit technique but needs a larger computer memory.

7.17.3 Finite Element Method

The finite element method enables an analysis to be made for complex shapes and for boundary conditions which are non-linear. This technique of numerical analysis originated in the work of engineers dealing with the problem of stress determination in an elastic continuum. The real continuum was divided into elements interconnected only at a finite number of nodal points where forces were introduced. Other than its applications to stress the method is now widely applied to a number of problems in process modelling, and references are given at the end of this Chapter.

Initially the volume to be analysed must be discretised, i.e. divided into elements, the meeting points of which are the "nodes". Within each element, the temperature is considered to be a linear function of the temperatures at the nodes. All individual relations are assembled into a global set of equations having the form

$$[H]\{T\} = \{F\}$$

$\{T\}$ is a column matrix of all nodal temperatures
$\{F\}$ is the thermal load vector contributed by the boundaries
$[H]$ is a stiffness matrix

The equations are solved using an iterative technique for $\{T\}$. A wide variety of boundary condtiions is used.

7.18 Artificial Intelligence/Expert Systems

Computer programs are ordinarily based on algorithms giving commands within a framework based on numerical solutions of mathematical equations. In place of the normal algorithm, artificial intelligence uses a logic system based on inference. The program is based on a knowledge base which may include data which can be inferred, i.e. it may not necessarily be exact, but can be used and tested in the programs. The program contains conditional statements. The logic system has also been termed "fuzzy". This allows computer programs to be written based on empirical knowledge generally covered by that attributed to experts. These programs are written in languages like LISP, PROLOG and C.

An Expert System is a particular type of computer programme of broad application in engineering and of much interest in process technology. As the fundamental, scientific knowledge of processes increases, it becomes possible to model processes and obtain exact solutions to the algorithms. However a large part of the knowledge of processes is empirical. This knowledge is known as ''heuristic'' and large areas of processes are understood by experts. This knowledge is largely based on experience, and very often obtained by trial and error. These systems normally require a computer scientist to write a programme using the knowledge of the expert. Because the way to expert knowledge is largely undefined, writing the programme, i.e., extracting logic from expertise requires special techniques. While the knowledge may be obtained from the expert, it may also be obtained by a programme which uses a learning process involving asking questions. These duplicate decision making techniques of the expert. Expert systems are available in a large number of process industries.

References

Bibel W, Petkoff J (Eds.) 1988 Artificial Intelligence. N. Holland.

Bolter JD (1986). Turing Man Western Culture in the Computer Age. Penguin Books.

Black JT (1988). The Design of Manufacturing Cells (Step One to Integrated Manufacturing Systems). Proc. Manuf. Int. Vol 3 p 143 Atlanta GA. ASME.

Brodie J, Murray JJ (1982). The Physics of Microfabrication. Plenum.

Chamberlin DA (1976). Relational Data Base Management Systems. ACM Computating Surveys 8:43–66.

Clark JP, Flemings MC (1986). Advanced Materials and the Economy Scientific American. 225(4) 43-49.

Delobel C, Adiba M (1985). Relational Data Base Systems, N.Holland.

Duckworth WE, Gear AE, Lockett AG (1977). A Guide to Operational Research 3rd Ed. Chapman Hall, London.

Groover MP (1980). Automation, Production Systems and Computer Aided Manufacturing. Prentice Hall.

Groover MP, Zimmers EW (1984). CAD, CAM Computer Aided Design and Manufacturing, Prentice Hall.

Handy C (1989). The Age of Unreason. Arrow Books, London.

Holland JR (1984). Flexible Manufacturing Systems 1st Ed.

Hunt VD (1983). Industrial Robotics Handbook. Industrial Press Inc. N.Y.

Kusiak A (Ed.). Flexible Manufacturing Systems. Methods and Studies. N. Holland.

Kenney RL (1982). Decision Analysis: An Overview. Operations Research 30, 5:803.

Kenney RL, Raiffa H (1976). Decisions with Multiple Objectives, Preferences and Value Tradeoffs. John Wiley N.Y.

Kochhar AK, Burns ND (1983). Microprocessors and their Manufacturing Applications. Edward Arnold.

Kowalik JS (1986). Knowledge Based Problem Solving. Prentice Hall.

Krockel H:, Reynard K, Stven G (Eds.) (1986). Factual Material Data Banks, CEC Workshop. Petten.

Miller WE, Automation of Metallurgical Processes. An Overview. Drive Systems Dept. GEC Salem.

Moffat DW (1987). Handbook of Manufacturing and Production, Management Formulas, Charts and Tables, Prentice Hall.

Muller RS, Kamins THI (1986). Device Electronics for Integrated Circuits 2nd Ed. John Wiley.

O'Brien JJ (1970). Management Information Systems. Van Nostrand Reinhold.

O'Shea T, Eisenstadt M (1984). Artificial Intelligence. Harper and Row.

Ott ER (1975). Process Quality Control McGraw-Hill.

Palm WJ (1986). Control Systems Engineering. John Wiley.

Parent M, Laurgeau C (1985). Robot Technology Vol. 5, Logic and Programming (English Transl.) Prentice Hall.

Pidd M (1984). Computer Simulation in Management Science, John Wiley.

Rydz JS (1986). Managing Innovation. Ballinger Publ Co.

Shingo S. (1985). A Revolution in Manufacturing. The SMED System. Productivity Press, Stanford.

Sol HJ, Takkenberg CAT, DeVries P (1985). Expert Systems and Artificial Intelligence in Decision Support Systems. Proc. 2nd MiniEuroConference, Luuton, Netherlands.

Tijnaelis D, McKee KE (1987). Manufacturing High Technology Handbook, Marcel Dekker.

Walker TC, Miller RK (Eds.) (1986). Expert Systems. An Assessment of Technology and Applications. SEAI Technical Publications, Madison GA 30650.

Zinkiewicz OC, Morgan K (1983). Finite Elements and Approximation, 1st Ed. John Wiley.

Zuboff S (1988). The Future of Work and Power. Basic Books, N.Y.

Subject Index